Nelson Maths

W0036064

3

This book belongs to:

_____

Workbook

Karen Morrison
Lisa Greenstein

OXFORD
UNIVERSITY PRESS

Great Clarendon Street, Oxford, OX2 6DP, United Kingdom

Oxford University Press is a department of the University of Oxford.

It furthers the University's objective of excellence in research, scholarship, and education by publishing worldwide. Oxford is a registered trade mark of Oxford University Press in the UK and in certain other countries.

First published 2022

British Library Cataloguing in Publication Data

Data available

ISBN: 978-1-382-01028-3

1 3 5 7 9 10 8 6 4 2

Paper used in the production of this book is a natural, recyclable product made from wood grown in sustainable forests. The manufacturing process conforms to the environmental regulations of the country of origin.

Printed in Great Britain by Bell and Bain Ltd, Glasgow

Acknowledgements

The publisher and authors would like to thank the following for permission to use photographs and other copyright material:

Cover: Matthieu Nivesse.

Artwork by Aviel Basil, Q2A Media, Pantek Media, and OKS Prepress.

Every effort has been made to contact copyright holders of material reproduced in this book. Any omissions will be rectified in subsequent printings if notice is given to the publisher.

Contents

Think maths

Mindset statements

Write or draw 'helpful' statements to help everyone in your class do well at maths.

Statements that help us to learn maths

I will try again.

➡ *Pupil Book page 7*

Number and place value

Work with tens

The picture shows 16 tens

The number is 160

We say: one hundred and sixty

Hundreds	Tens	Ones
1	6	0

1 How many tens are there? Write how many tens, and the number.

a _____ tens

b _____ tens

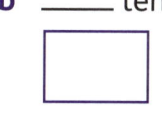

c _____ tens

d _____ tens

H	T	O
4	7	0

e _____ tens

H	T	O
9	8	0

f _____ tens

H	T	O
7	3	0

➡ Pupil Book page 11

Place value

1 Complete the table. The first row is done for you.

Place-value cards	Hundreds, tens and ones	Numeral and number name
1 4 3	1 hundred 4 tens 3 ones	143 one hundred and forty-three
2 3 2	☐ hundreds ☐ tens ☐ ones	
5 1 0	☐ hundreds ☐ tens ☐ ones	
6 8 1	☐ hundreds ☐ tens ☐ ones	

2 Add up the hundreds, tens and ones. Write the total.

a 100 10 10 1 ☐

b 100 100 1
1 1 ☐

c 100 100
100 100 1
100 100 1
100 10 1 ☐

d 100 100 100
10 10 10
1 1 10
1 1 1 ☐

➡ *Pupil Book page 12*

More place value

H	T	O
••	:::	:•
	253	

H	T	O
•	:::	:•
	153	

These place-value tables use dots to show numbers of hundreds (H), tens (T) and ones (O). Also, remember that:

253 > 153

153 < 253

1 Draw dots to show each number.

Write < or > to complete each number sentence.

a

H	T	O
	117	

H	T	O
	137	

117 ◯ 137

137 ◯ 117

b

H	T	O
	530	

H	T	O
	503	

530 ◯ 503

503 ◯ 530

c

H	T	O
	927	

H	T	O
	972	

927 ◯ 972

972 ◯ 927

2 Complete the additions.

a $500 + 200 + 300 =$ ☐   **b** $100 + 100 + 800 =$ ☐   **c** $400 + 400 + 200 =$ ☐

3 Write three different additions that each have a sum of 900.

_____

_____

_____

➡ *Pupil Book page 12*

Spot the mistake

1 Look at the underlined digit in each number. Some of the values given for the underlined digits are wrong.

Put a tick (✔) next to the correct values. If the value is wrong, write the correct value. The first one is done for you.

a

☐ 200   20

b

☐ 2   ☐

c

☐ 80

d

☐ 400   ☐

e

☐ 900   ☐

f

☐ 70

g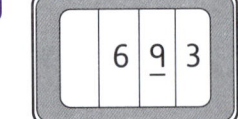

☐ 90   ☐

h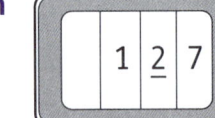

☐ 20   ☐

i

☐ 80

j

☐ 3   ☐

k

☐ 500   ☐

l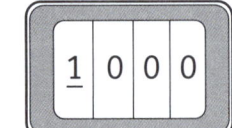

☐ 100

---

2 Draw beads to show the hundreds, tens and ones for these numbers.

a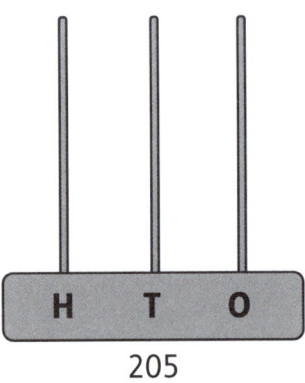

205

b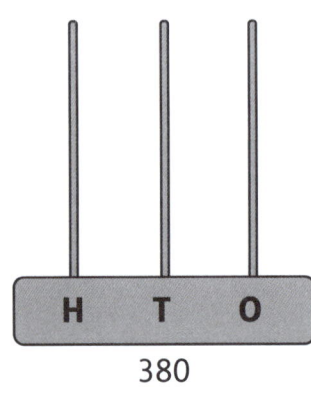

380

c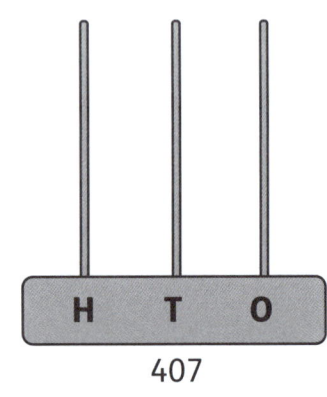

407

➡ *Pupil Book page 13*

Partitioning

When we partition a number into hundreds, tens and ones, we make a **regular partition**. When we partition the number in other ways, the partition is **irregular**.

Regular partition
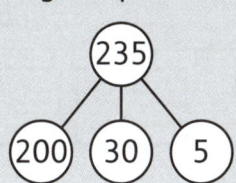

Irregular partition
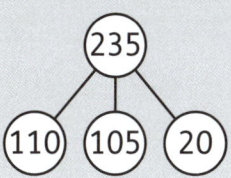

1 Show two more ways to partition 235.

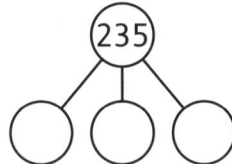 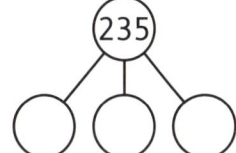

2 Partition these numbers using regular partitioning (hundreds, tens and ones).

a 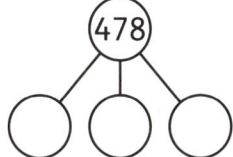 b 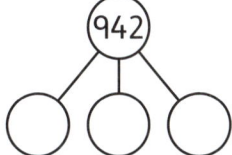 c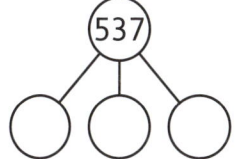

3 Choose your own 3-digit number. Partition it in three different ways.

a 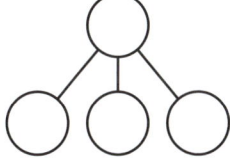 b 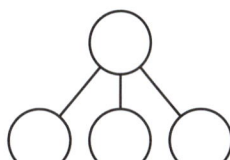 c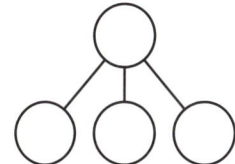

4 David added up these counters to 243.
What was his mistake?

_____

10	100	1
10	100	1
	100	1
	100	

5 Fill in the missing numbers.

a $500 + 90 + 5 = \boxed{}$ b $3 + 40 + 500 + = \boxed{}$ c $80 + 4 + 900 = \boxed{}$

d $583 = 500 + \boxed{} + 3$ e $479 = 70 + \boxed{} + 9$ f $794 - 90 = \boxed{}$

g $872 - \boxed{} - 2 = 800$ h $\boxed{} = 407 + 40$ i $\boxed{} = 360 + 8$

➡ Pupil Book page 14

Number lines

1 Fill in the missing numbers on each number line.

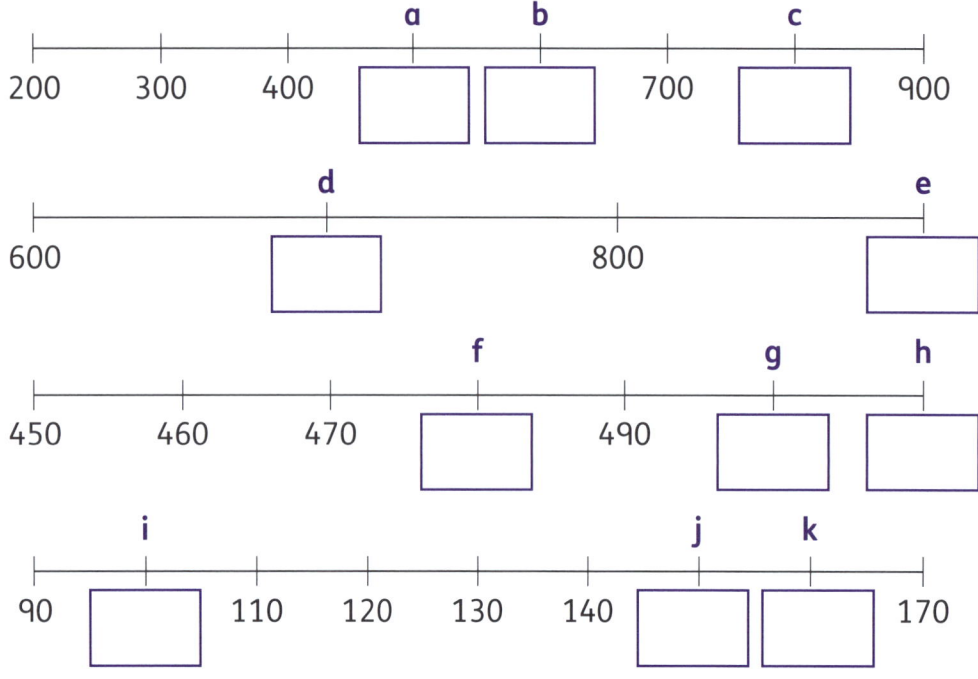

2 Only two numbers are shown on each number line. Write what you think the other marks represent. Tell your partner how you worked it out.

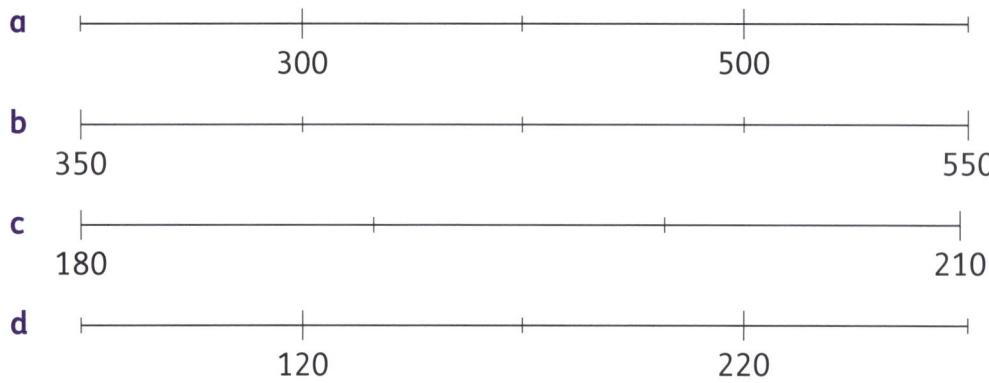

3 Estimate the value of the missing numbers. Write your estimates in the boxes.

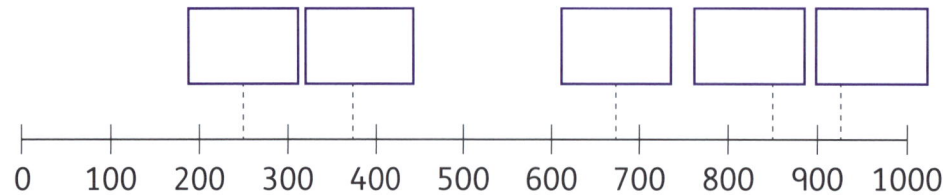

4 Estimate where each number should go on the number line. Write each number on the number line.

450	600	100	879	25

0   1000

➡ *Pupil Book page 15*

Order numbers

1 Colour the greatest number in each set.

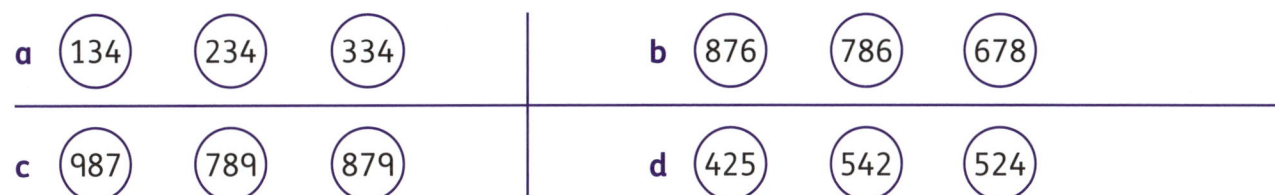

a (134) (234) (334)   b (876) (786) (678)

c (987) (789) (879)   d (425) (542) (524)

2 Colour the smallest number in each set.

a (432) (342) (234)   b (564) (456) (645)

c (231) (123) (132)   d (279) (297) (207)

3 Write the numbers in order from smallest to greatest.

| 179 | 366 | 529 | 107 | 201 | 507 | 963 |

☐ ☐ ☐ ☐ ☐ ☐ ☐

4 Write the numbers in order from greatest to smallest.

| 987 | 900 | 978 | 943 | 956 | 999 | 988 |

☐ ☐ ☐ ☐ ☐ ☐ ☐

5 This is a 200 to 300 number line marked in tens.

200   300

Write these numbers in the correct positions on the number line.

| 290 | 250 | 280 | 210 | 260 |

➡ *Pupil Book page 16*

Compare numbers

1 Use the three digits.

Write the greatest number you can make.

Write the smallest number you can make.
The one opposite is done for you.

7
5
1

H	T	O
7	5	1

H	T	O
1	5	7

a

H	T	O

H	T	O

b

1
8
5

H	T	O

H	T	O

c

7
6
1

H	T	O

H	T	O

d

7
9
8

H	T	O

H	T	O

e

9
5
8

H	T	O

H	T	O

f

3
4
7

H	T	O

H	T	O

g

H	T	O

H	T	O

h

5
1
1

H	T	O

H	T	O

2 Fill in the missing numbers.

a 100 less   100 more

☐ ← **700** → ☐

b 10 less   10 more

☐ ← **430** → ☐

c 100 less   100 more

☐ ← **635** → ☐

d 10 less   10 more

☐ ← **921** → ☐

➡ *Pupil Book page 17*

Round numbers

1 Which number is closest to the number in the speech bubble? Circle the answer.

a
I ate about 200 mosquitoes.

146   198   250

b
I ate about 300 mosquitoes.

210   249   328

c
I ate about 100 mosquitoes.

106   160   201

d
I ate about 500 mosquitoes.

210   475   610

e
I ate about 400 mosquitoes.

398   456   327

f
I ate about 300 mosquitoes.

249   376   260

g
I ate about 100 mosquitoes.

50   155   36

h
I ate about 300 mosquitoes.

187   256   301

i
I ate about 500 mosquitoes.

662   559   530

➡ *Pupil Book page 19*

Measure paths

1 How long is each path? First measure each section to the nearest cm. Then add.

a

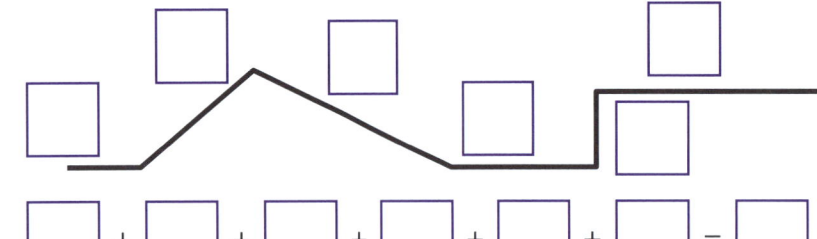

☐ + ☐ + ☐ + ☐ + ☐ + ☐ = ☐

b

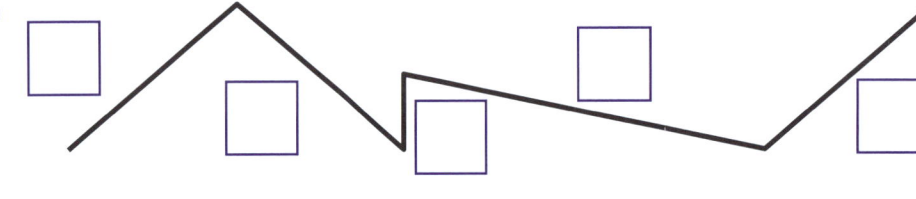

☐ + ☐ + ☐ + ☐ + ☐ = ☐

c

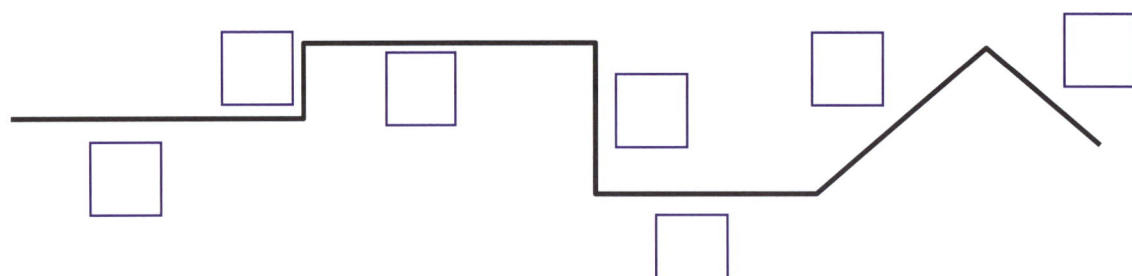

☐ + ☐ + ☐ + ☐ + ☐ + ☐ + ☐ = ☐

2 Draw a path with 3 sections, with a total length of 15 cm. How many different ways can you do it? Compare with a partner.

➡️ *Pupil Book page 21*

Estimate and measure in metres

1 Estimate the length or height of each object in metres. Write your estimate in the table. Then measure each object to the nearest half-metre. Write your measurements.

Choose two more objects. Estimate and then measure their height or length.

Object	Estimate	Measure
Teacher's table	☐ metres	☐ metres
Noticeboard	☐ metres	☐ metres
Classroom wall	☐ metres	☐ metres
Door	☐ metres	☐ metres
_____	☐ metres	☐ metres
_____	☐ metres	☐ metres

2 Choose objects in your school to measure.

Estimate and then measure the width and the length of each object. Work out the difference between your estimate and your measurement.

What I measured	Estimate	Measure	Difference

3 Estimate the length of the shorter line in each pair by comparing it to the 10 m line. Write your estimate next to the line.

a   10 m

b   10 m

c   10 m

d   10 m

➡ *Pupil Book page 23*

Patterns in space

1 Fill in the gaps in each sequence. Draw the shapes and write the numbers.

a

2   4   ☐   8   ☐

b

 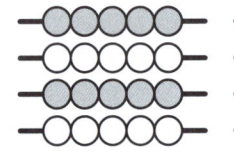

5   ☐   15   20   25

c

☐   6   9   ☐   15

d

4   ☐   12   ☐   20

2 Make your own patterns. Use shapes and numbers.

a

b

➡ *Pupil Book page 24*

Function machines

A function machine changes the number you put in according to a **rule**. The new number comes out the other side.

The numbers you put in are the inputs.

The numbers that come out are the outputs.

The rule for this function machine is 'add 5'.

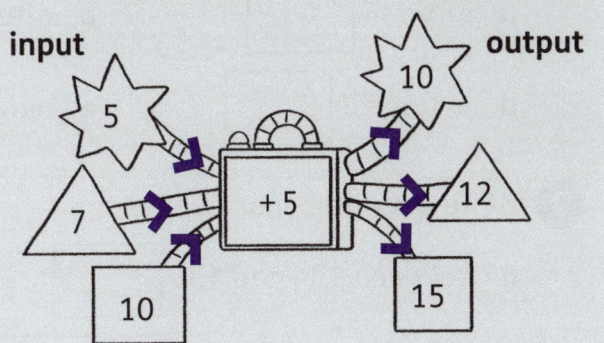

1 Fill in the missing numbers.

a

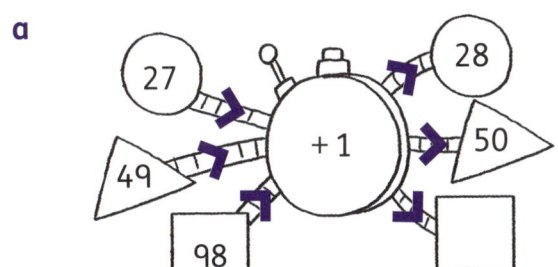

b

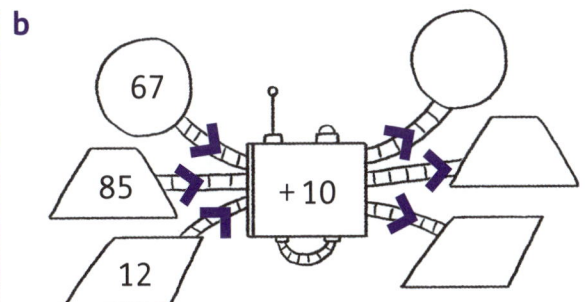

c

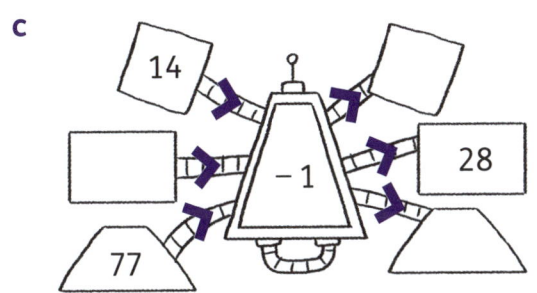

d

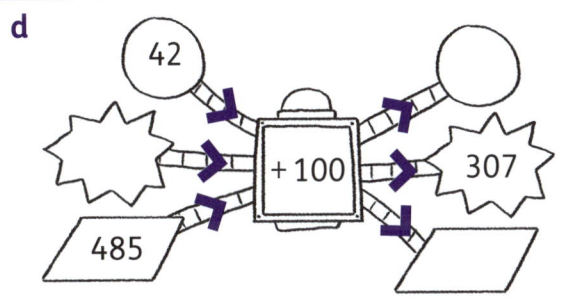

e

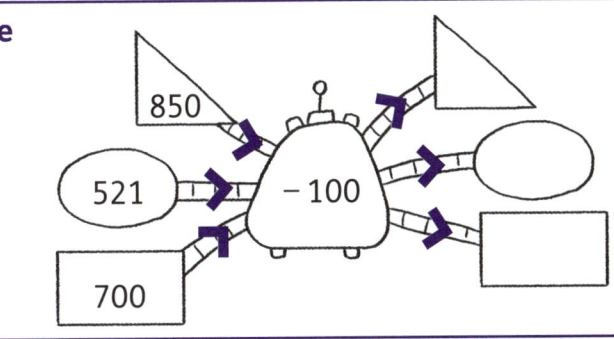

f

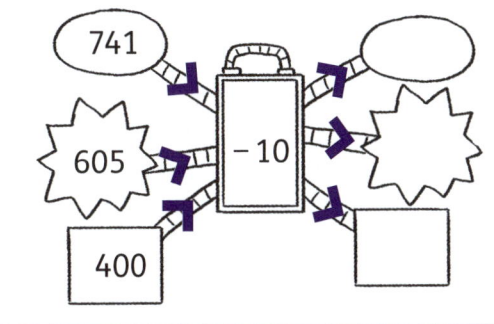

g

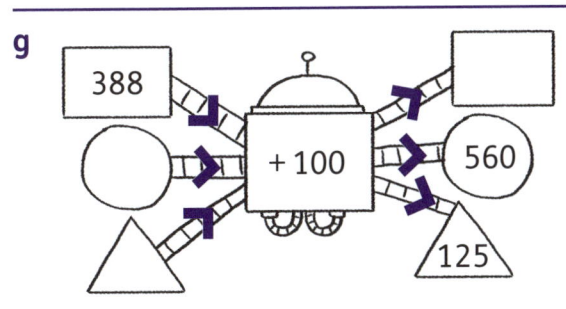

h

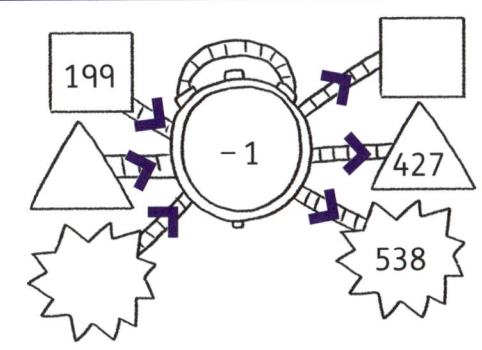

➡ *Pupil Book page 25*

Missing numbers

1 Write the number:

a after 34 ☐   **b** after 81 ☐   **c** after 99 ☐

d before 27 ☐   **e** before 88 ☐   **f** before 40 ☐

2 Write the number or numbers:

a between 134 and 136 ☐   **b** between 269 and 271 ☐

c between 366 and 370 ☐

3 Write the number that is 1 more than:

a 335 ☐   **b** 450 ☐   **c** 639 ☐   **d** 789 ☐

4 Write the number that is 10 more than:

a 135 ☐   **b** 250 ☐   **c** 339 ☐   **d** 489 ☐

5 Write the number that is 1 less than:

a 991 ☐   **b** 845 ☐   **c** 700 ☐   **d** 673 ☐

6 Write the number that is 10 less than:

a 291 ☐   **b** 345 ☐   **c** 700 ☐   **d** 673 ☐

7 Write the number name for the number that is:

a one more than fifty _____

b a hundred more than forty _____

c ten more than four hundred _____

d a hundred less than eight hundred _____

8 Write the missing numbers.

a 500 ⟶ 100 more is ☐   **b** ☐ ⟶ 100 less is 600

c 340 ⟶ 100 more is ☐   **d** ☐ ⟶ 100 less is 450

e 990 ⟶ 10 more is ☐   **f** ☐ ⟶ 10 less is 125

➡ *Pupil Book page 25*

Counting patterns to 1000

1 Write the number that is 1 more than each number. The first one is done for you.

   a four hundred and twenty-three `424`   **b** two hundred and sixteen [ ]

   c five hundred and fifty [ ]   **d** six hundred and eighty-two [ ]

   e nine hundred and thirty-five [ ]   **f** seven hundred and four [ ]

2 To get from 150 to 160, you could count on in tens.

   How would you count to get from the first number to the second number in each pair? The first one is done for you.

   a 647, 654 _on in sevens_   **b** 500, 501 _____

   c 836, 835 _____   **d** 942, 952 _____

   e 750, 740 _____   **f** 429, 430 _____

   g 316, 416 _____   **h** 609, 599 _____

3 Write the number that is 10 more than each number.

   a 655 [ ]   **b** 399 [ ]   **c** 999 [ ]

4 Write the number that is 10 less than each number.

   a 875 [ ]   **b** 600 [ ]   **c** 749 [ ]

5 Write the number that is 100 more than each number.

   a 200 [ ]   **b** 139 [ ]   **c** 650 [ ]

6 Write the number that is 100 less than each number.

   a 500 [ ]   **b** 550 [ ]   **c** 490 [ ]

7 Would you count on or back from the first number to the second number? Circle the correct answer.

   a 397 ⟶ 379   on   back   **b** 641 ⟶ 614   on   back

   c 534 ⟶ 543   on   back   **d** 650 ⟶ 615   on   back

   e 837 ⟶ 873   on   back   **f** 768 ⟶ 786   on   back

➡️ *Pupil Book page 26*

Right angles

1 Mark all the right angles on each path. Write how many right angles there are.

a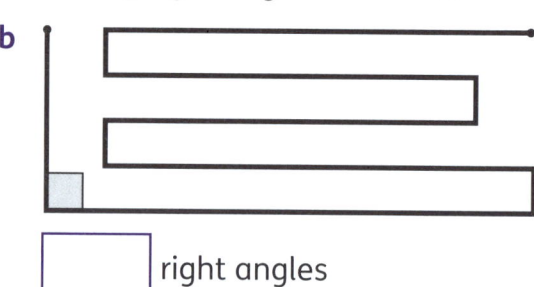

[ ] right angles

b

[ ] right angles

2 Write the name of each shape. Write how many right angles there are in each shape.

a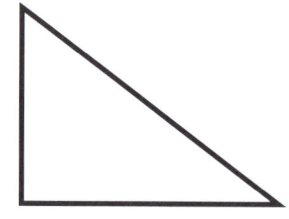

_____

[ ] right angles

b

_____

[ ] right angles

c

_____

[ ] right angles

d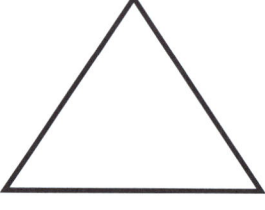

_____

[ ] right angles

e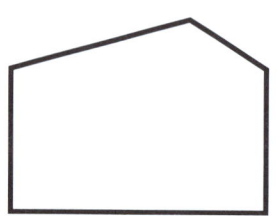

_____

[ ] right angles

f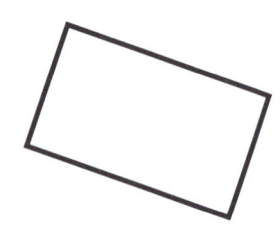

_____

[ ] right angles

3 Draw three different shapes that each have one or more right angles.

➡ *Pupil Book page 29*

Angles, lines and turns

1

a	Draw 2 right angles in different positions.

b	Draw 2 pairs of perpendicular lines.

c	Draw some sets of parallel lines.

d	Fill this box with horizontal lines.

e	Fill this box with vertical lines.

f	Draw a half turn.

g	Draw a $\frac{3}{4}$ turn.

h	Draw a shape and label the kinds of angles and lines it has.

➡ *Pupil Book page 32*

Polygons

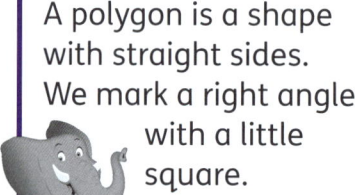

A polygon is a shape with straight sides. We mark a right angle with a little square.

Shapes and angles

1 Mark all the right angles in these shapes.

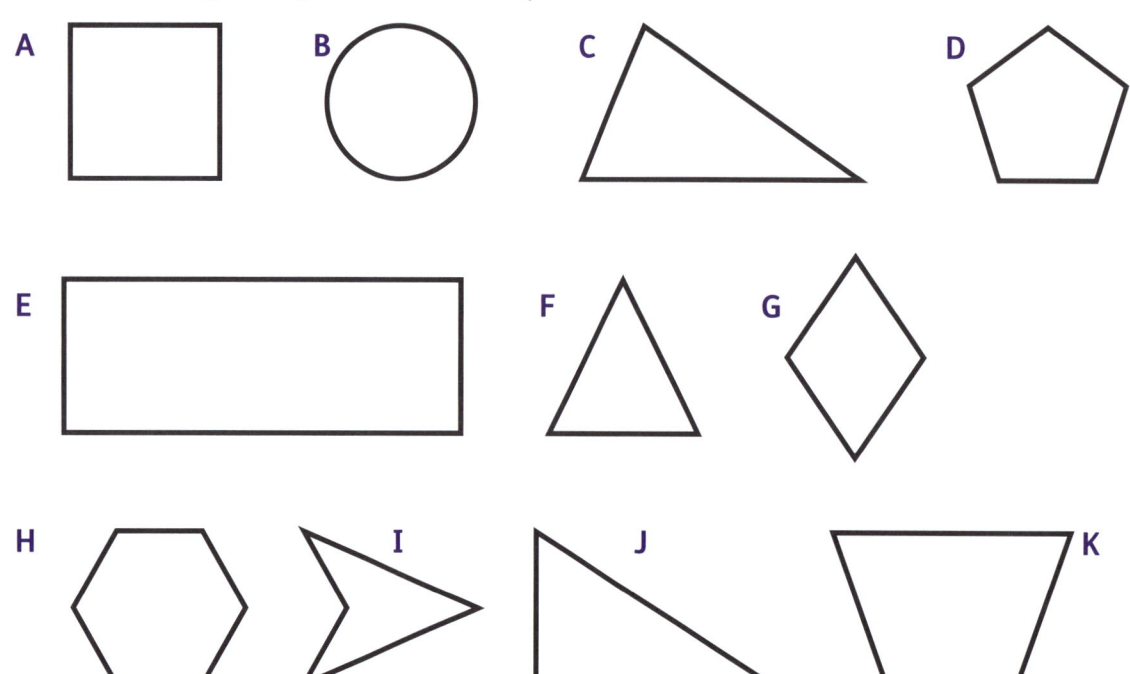

A B C D

E F G

H I J K

2 Write the letters of the shapes from question 1 in the correct rows of this table.

	Shape
No right angles	
1 right angle	
More than 1 right angle	

3 **a** Write the letters of the polygons from question 1 in the correct columns of this table.

Regular polygons	Irregular polygons

b One of the shapes does not belong in the polygon table at all.

Which shape is it? _____

Explain why: _____

_____

_____

➡ *Pupil Book page 34*

Sketch 2D shapes

1 Use a ruler and a set square if you have one.

a	Sketch two different right-angled triangles.

b	Sketch a rectangle with two sides that are 4 cm long and two sides that are 2 cm long.

c	Sketch a square with sides that are 3 cm long.

d	Sketch an irregular pentagon with two right angles.

➡ *Pupil Book page 36*

Alphabet symmetry

1 Tick (✔) the letters that are symmetrical. Draw a line of symmetry on the letters that are symmetrical.

A B C D E F

G H I J K L

M N O P Q R

S T U V W X

Y Z

2 The word WOW is symmetrical if we draw a mirror line vertically through the O. What is the longest word you can make that is symmetrical?

➡ *Pupil Book page 37*

Symmetry

1 Complete each shape to make a shape that is symmetrical about the dashed line.

a
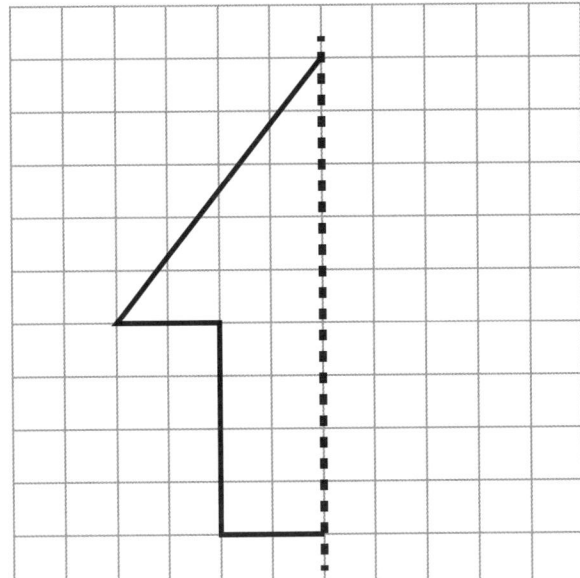

b
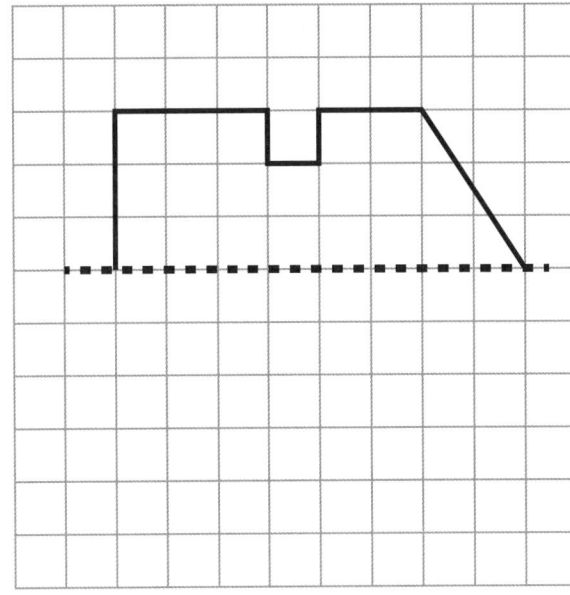

2 Shade two more squares in each grid to make a symmetrical block pattern.

a
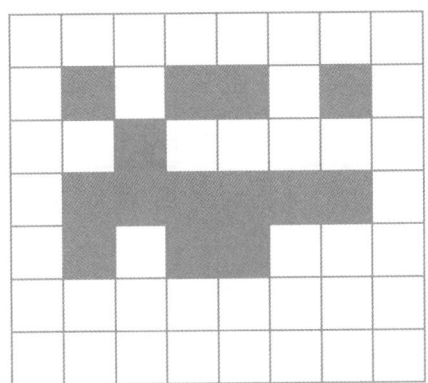

b
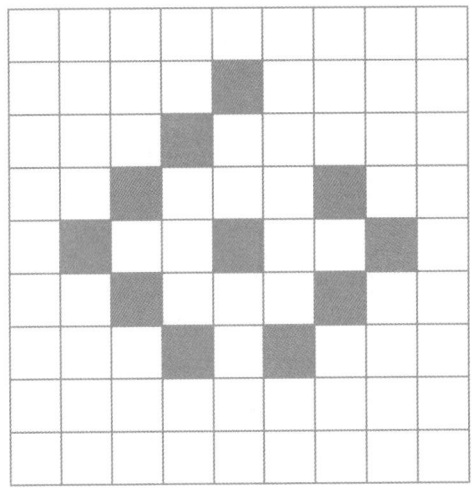

3 Draw a symmetrical block pattern of your own.

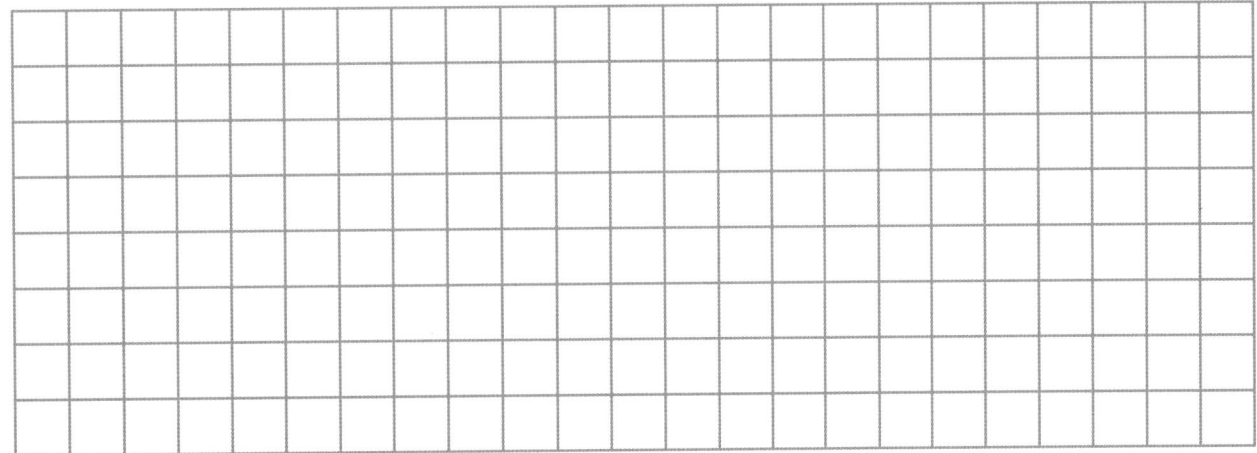

➡ *Pupil Book page 37*

Revise number facts

1 Fill in the missing numbers in these function machines.

a
2, 3, 7 → +10 → ( ), ( ), ( )

b
12, 23, 47 → −10 → ( ), ( ), ( )

c
7, 8, 9 → [ ] → 12, 13, 14

d
9, 16, 8 → [ ] → 4, 11, 3

e
9, 4, 7 → [ ] → 17, 12, 15

f
4, 11, 23 → −4 → ( ), ( ), ( )

g
( ), ( ), ( ) → +6 → 14, 27, 31

h
( ), ( ), ( ) → −9 → 7, 3, 11

2 Draw two diagrams or function machines. Show what happens to three different numbers with these rules.

a   add 100

b   subtract 100

➡️ *Pupil Book page 41*

Make 100

10	20	30	40	50	60	70	80	90	100

1 Write the missing numbers.

a 10 + ☐ = 100

b 25 + ☐ = 100

c 40 + ☐ = 100

d 35 + ☐ = 100

e 50 + ☐ = 100

f 65 + ☐ = 100

g 30 + ☐ = 100

h 45 + ☐ = 100

i 80 + ☐ = 100

j 95 + ☐ = 100

2 Write the answers.

a 100 – 90 = ☐

b 100 – 15 = ☐

c 100 – 80 = ☐

d 100 – 25 = ☐

e 100 – 70 = ☐

f 100 – 35 = ☐

g 100 – 60 = ☐

h 100 – 45 = ☐

i 100 – 50 = ☐

j 100 – 65 = ☐

3 Colour pairs of numbers that make 100. Use a different colour for each pair.

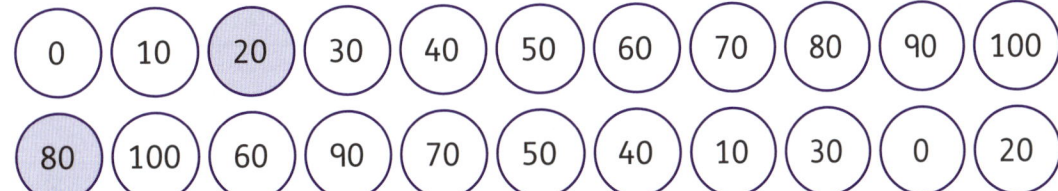

4 How many steps of 100 would you take forwards or backwards on a number line to get from the first number to the second number in each pair?

a 600 to 100 ☐ steps

b 100 to 1000 ☐ steps

c 1000 to 0 ☐ steps

d 1000 to 500 ☐ steps

e 225 to 525 ☐ steps

f 744 to 444 ☐ steps

▶ Pupil Book page 42

Add to 100 and 200

1 Complete the number sentences.

a 60 + 40 = ☐

b 30 + ☐ = 100

c 50 + ☐ = 100

d 100 = 20 + ☐

---

2 Fill in the missing number in each part–whole diagram.

a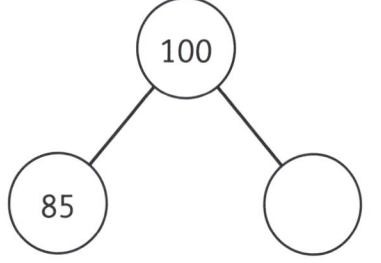

b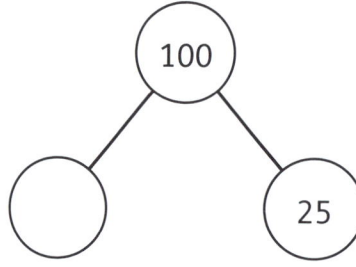

c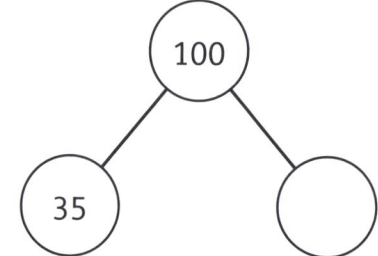

d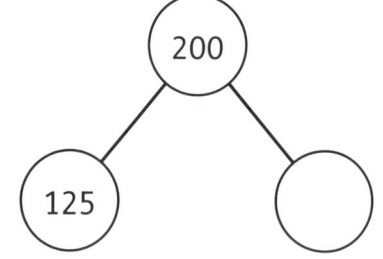

e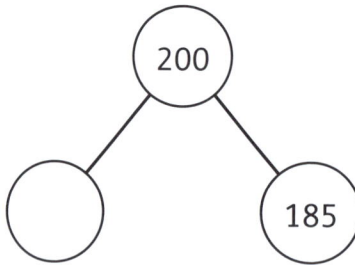

f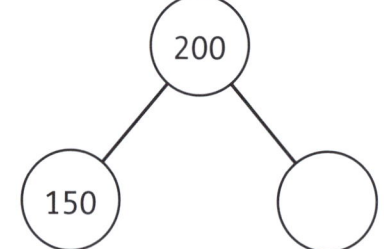

---

3 Make pairs of numbers that total 100.

a 42 + ☐

b 67 + ☐

c ☐ + 53

d ☐ + 19

e ☐ + 74

f 81 + ☐

---

4 Circle the pairs that make 100.

38 + 72

63 + 47

49 + 51

24 + 86

89 + 11

33 + 67

➡ *Pupil Book page 43*

Add in columns

1 Find the totals.

a
```
  T  O
  2  3
+ 1  4
───────
```

b
```
  T  O
  1  3
+ 4  2
───────
```

c
```
  T  O
  5  8
+ 1  1
───────
```

d
```
  T  O
  1  1
  2  3
+ 4  2
───────
```

e
```
  T  O
  6  3
  3  1
+    5
───────
```

f
```
  T  O
  4  4
  1  3
+ 2  2
───────
```

2 Add.

a
```
  H  T  O
  1  2  4
+ 3  0  1
──────────
```

b
```
  H  T  O
  6  1  7
+ 1  2  1
──────────
```

c
```
  H  T  O
  4  7  6
+ 3  2  3
──────────
```

3 Fill in the missing digits to make each addition correct.

a
```
   1   3   1
+  2  [ ]  3
────────────
   3   7   4
```

b
```
  [ ]   1   4
+  1    3  [ ]
─────────────
   8    4   9
```

c
```
   4    0  [ ]
+ [ ]   4   2
─────────────
   7    4   2
```

4 Lucia sells 103 orange juices, 210 lemonades and 114 coconut waters. How many drinks does she sell altogether?

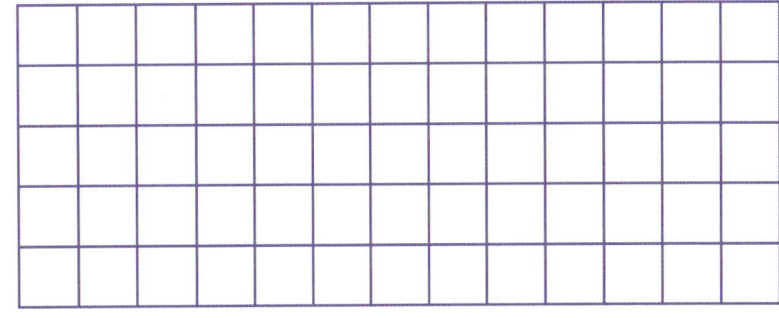

5 Add in columns to check that this bar model is correct.

387	
265	122

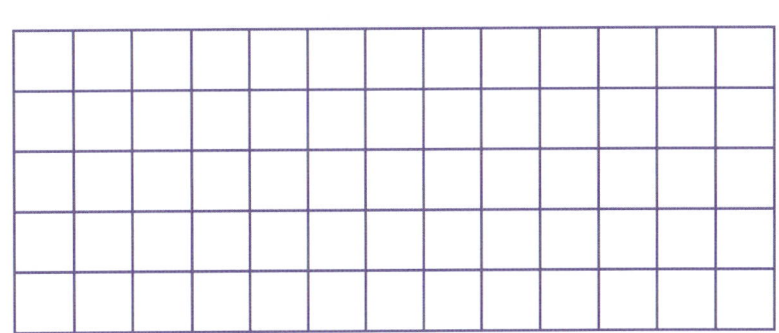

➡ *Pupil Book page 44 and page 45*

Add and subtract

1 Draw arrows on the number line to help you add or subtract.

a 78 + 40 = ▢

```
|----|----|----|----|----|----|----|----|----|----|
58   68   78   88   98   108  118  128  138  148
```

b 193 – 60 = ▢

```
|----|----|----|----|----|----|----|----|----|----|
113  123  133  143  153  163  173  183  193  203  213
```

c 792 + 30 = ▢

```
|----|----|----|----|----|----|----|----|----|
772  782   792   802   812   822   832   842   852
```

d 515 – 70 = ▢

```
|----|----|----|----|----|----|----|----|----|
435  445  455  465  475  485  495  505  515  525
```

2 Expand the numbers to add or subtract. The first one is done for you.

a 632 – 21

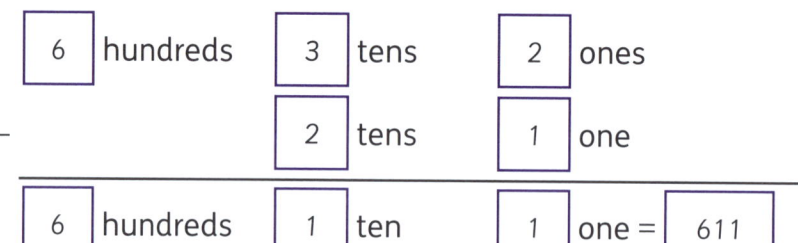

	6 hundreds	3 tens	2 ones
–		2 tens	1 one
	6 hundreds	1 ten	1 one = 611

b 548 – 35

	▢ hundreds	▢ tens	▢ ones
–		▢ tens	▢ ones
	▢ hundreds	▢ tens	▢ ones = ▢

c 756 + 33

	▢ hundreds	▢ tens	▢ ones
+		▢ tens	▢ ones
	▢ hundreds	▢ tens	▢ ones = ▢

➡ *Pupil Book page 47 and page 48*

Addition with regrouping

1 Add.

a
```
  H  T  O
  1  4  4
+     7  2
_____
```

b
```
  H  T  O
  2  3  7
+ 1  1  9
_____
```

c
```
  H  T  O
  5  1  9
+     9  9
_____
```

d
```
  H  T  O
  5  7  9
  1  0  3
+ 2  1  4
_____
```

e
```
  H  T  O
  6  1  5
  1  4  9
+ 2  1  6
_____
```

f
```
  H  T  O
  3  0  3
     1  8
+ 4  8  1
_____
```

2 Add in columns to check whether these bar models are correct.

a

793		
315	464	315

b

685		
118	541	26

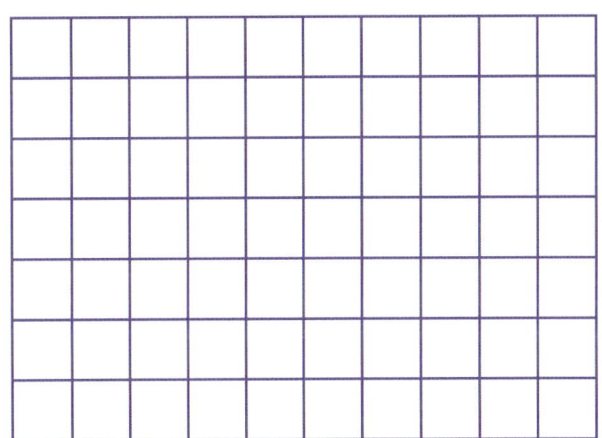

3 Jabu posted a video on a social media site. His video got 113 views on Monday, 247 views on Tuesday and 317 views on Wednesday. How many views did he get altogether over the three days?

➡ *Pupil Book page 49*

Subtract in columns

1 Subtract.

a
```
  T  O
  8  9
- 1  3
_____
```

b
```
  T  O
  4  6
- 3  0
_____
```

c
```
  T  O
  7  5
- 4  3
_____
```

d
```
  H  T  O
  7  1  8
- 3  0  4
_____
```

e
```
  H  T  O
  5  7  9
- 2  3  5
_____
```

f
```
  H  T  O
  6  2  8
- 3  2  4
_____
```

2 Fill in the missing numbers to make each subtraction correct.

a
```
  7  3  7
- 1  □  6
_____
  6  1  1
```

b
```
  8  9  9
- 4  1  □
_____
  4  8  2
```

c
```
  6  1  3
- □  0  2
_____
  3  1  1
```

d
```
  □  0  6
- 4  0  3
_____
  3  0  3
```

e
```
  9  □  5
- 4  9  2
_____
  5  0  3
```

f
```
  7  4  □
- □  0  4
_____
  4  4  1
```

3 Subtract in columns to help you solve the problems.

a Lovena has 685 ml of juice. She drinks 325 ml. How much is left?

b Craig's house is 878 m from the shop. He runs 414 m, then stops. How much further must he go to reach the shop?

➡ *Pupil Book page 50*

Subtraction with regrouping

1 Subtract.

a
```
    3  7
  - 1  9
  _____
```

b
```
    4  5
  - 2  8
  _____
```

c
```
    6  6
  - 3  7
  _____
```

2 Find the difference.

a
```
    9  4  4
  -    2  7
  _____
```

b
```
    6  7  5
  - 1  4  9
  _____
```

c
```
    5  2  3
  - 1  1  5
  _____
```

3 Year 3 collects 284 glass bottles. They send 155 bottles to the art room and they send the rest for recycling. How many bottles do they send for recycling?

4 Complete these part–whole diagrams. You can use the space underneath to work out the answers using columns.

a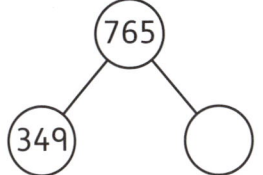
b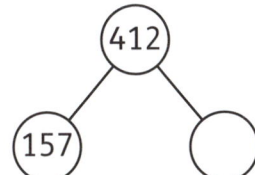
c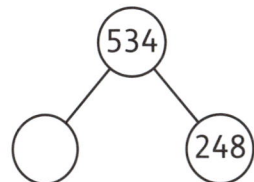

➡ *Pupil Book page 50*

Money

Money amounts

1 Draw notes and coins to show how you would pay for each item.

Item and price	Coins and notes I could use
a CASHEWS $4·50	
b WAX CRAYONS $8·50	
c $15·60	
d $12·80 DIARY	
e $13·70	

2 How much change do you get from $10 if you buy:

a the cashews [ ]   b the crayons [ ]

3 How much change do you get from $20 if you buy:

a the calculator [ ]   b the diary [ ]

c the cap [ ]

➡ *Pupil Book page 52 and page 53*

Work with money

1p   2p   5p   10p   20p   50p

Remember, these are some of the coins used in the UK. One pound (£1) = one hundred pence (100p).

1 Look at the amount you already have. Write values on the coins to make a total of £1.00. Use as few coins as possible.

Amount I have	Coins I need to make £1.00
25p	◯ ◯ ◯ ◯ ◯
75p	◯ ◯ ◯ ◯ ◯
52p	◯ ◯ ◯ ◯ ◯
88p	◯ ◯ ◯ ◯ ◯
90p	◯ ◯ ◯ ◯ ◯
18p	◯ ◯ ◯ ◯ ◯
39p	◯ ◯ ◯ ◯ ◯

2 How much do two of each item cost?

a 45p [   ]   b 90p [   ]   c 80p [   ]

d 85p [   ]   e £1.50 [   ]   f £3.50 [   ]

3 How much change will you get from £2 if you spend each amount?

a £1.25 [   ]   b £1.45 [   ]   c 50p [   ]

d 10p [   ]   e 99p [   ]   f £1.73 [   ]

➡ *Pupil Book page 54*

Mass

Things in my kitchen

1 Draw a picture of a kitchen scale.

   a How does your scale show the mass? Does it have a pointer or does it have an LCD (electronic panel)?

      _____

   b What is the heaviest mass your scale can measure?

      ┌─────────────────────────┐
      │   │
      └─────────────────────────┘

   c Does your scale need batteries? Why or why not?

      _____

2 Choose six different objects to measure on your scale. Write the name of each object. Estimate and then weigh to find the actual mass.

Object	Estimate	Actual mass

3 Write the masses from question 2 in order from lightest to heaviest.

_____

_____

➡ Pupil Book page 55

Read scales

1 Read the scales. Write each mass in two different ways.

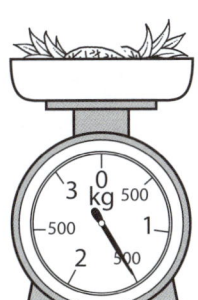

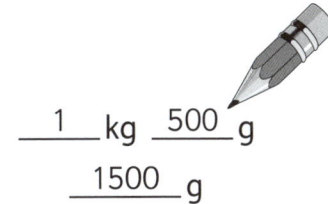

___1___ kg ___500___ g

___1500___ g

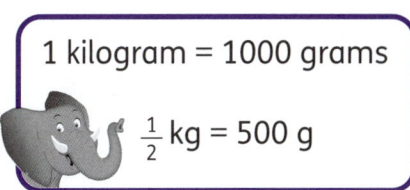

1 kilogram = 1000 grams

$\frac{1}{2}$ kg = 500 g

a

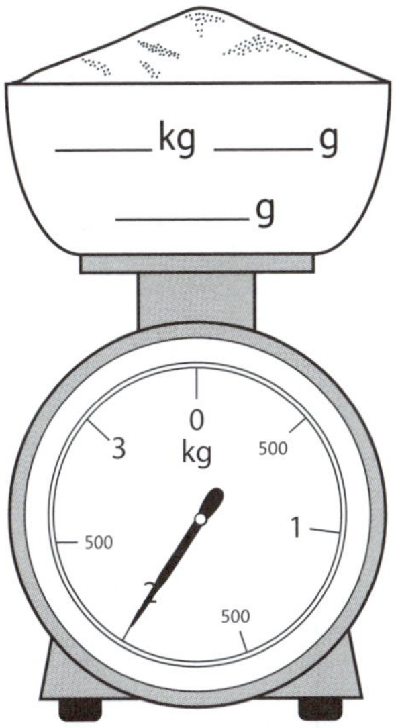

_____ kg _____ g

_____ g

b

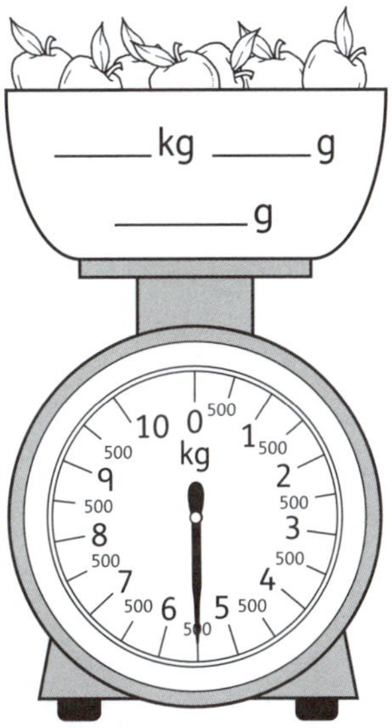

_____ kg _____ g

_____ g

c

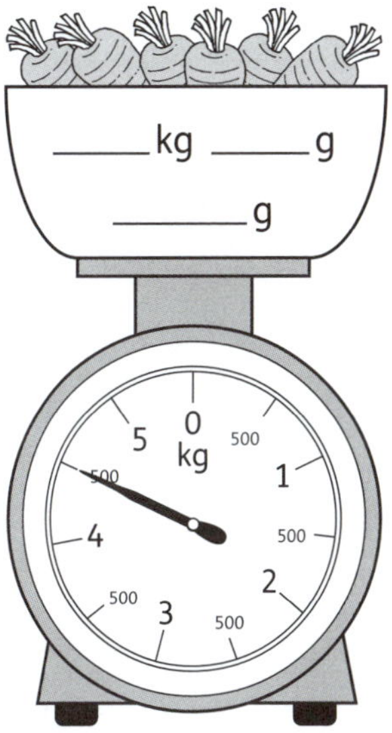

_____ kg _____ g

_____ g

d

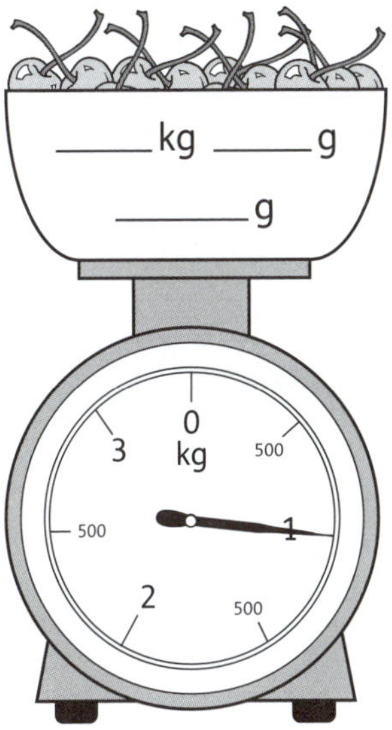

_____ kg _____ g

_____ g

e

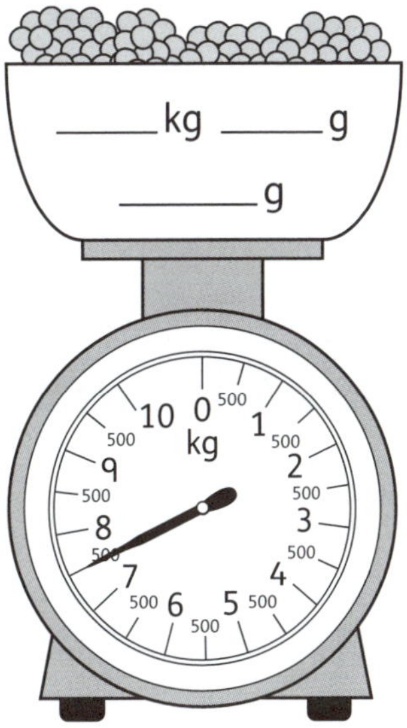

_____ kg _____ g

_____ g

f

_____ kg _____ g

_____ g

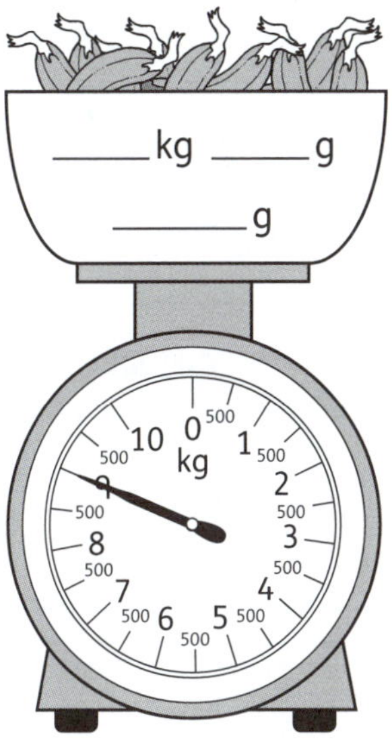

2 Some of the markings are missing on these scales. Estimate the mass shown by the pointer.

a

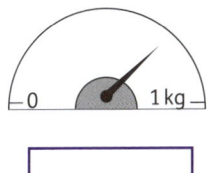

b

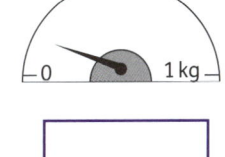

c

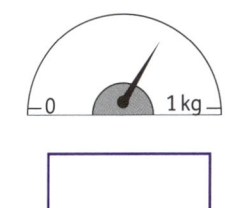

➤ *Pupil Book page 57*

More about mass

We sometimes use the word 'weight' to mean mass.

1 The school nurse has weighed nine children. Read the scale and write each mass to the nearest kilogram.

a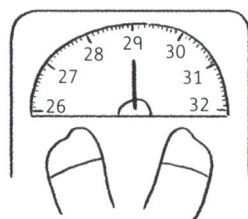

_____ kg

b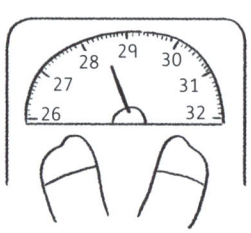

_____ kg

c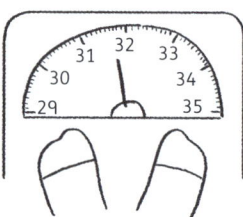

_____ kg

d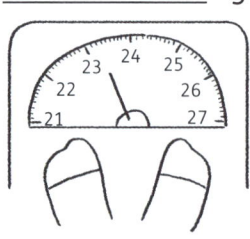

_____ kg

e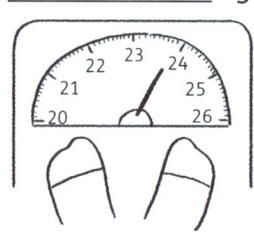

_____ kg

f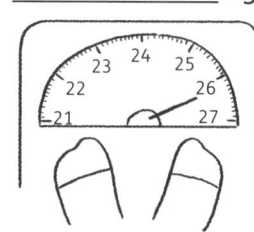

_____ kg

g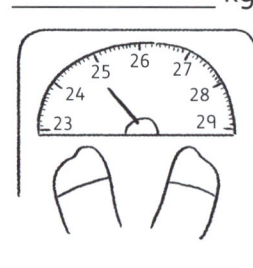

_____ kg

h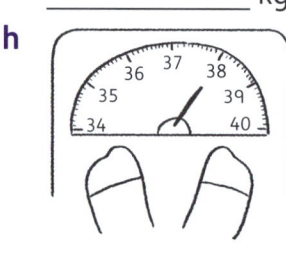

_____ kg

i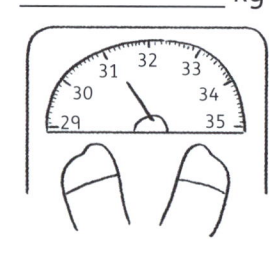

_____ kg

2 A post office gives this guide for the maximum mass and size of parcels:

Parcel size	Maximum mass	Maximum length	Maximum width	Maximum height
Small	2 kg	45 cm	35 cm	16 cm
Medium	20 kg	61 cm	46 cm	46 cm
Large	30 kg	1.5 m	1 m	50 cm

a I want to send a parcel that weighs 17 kg.
What is the longest it is allowed to be?

b I want to send a parcel that is 19 kg and 1.2 m long.
Is my parcel medium or large?

_____

➡ *Pupil Book page 57*

Multiplication and division

Revise multiplication and division

1 Complete the number sentences.

a

$5 \times 2 = \boxed{}$

$10 \div 2 = \boxed{}$

$2 \times 5 = \boxed{}$

$10 \div 5 = \boxed{}$

b

$2 \times 4 = \boxed{}$

$8 \div 2 = \boxed{}$

$4 \times 2 = \boxed{}$

$8 \div 4 = \boxed{}$

c

$4 \times 3 = \boxed{}$

$12 \div 3 = \boxed{}$

$3 \times 4 = \boxed{}$

$12 \div \boxed{} = 3$

d

$4 \times 5 = \boxed{}$

$20 \div \boxed{} = 5$

$5 \times 4 = \boxed{}$

$20 \div 5 = \boxed{}$

e

$10 \times 2 = \boxed{}$

$20 \div 2 = \boxed{}$

$2 \times 10 = \boxed{}$

$20 \div 10 = \boxed{}$

f

$8 \times 2 = \boxed{}$

$16 \div \boxed{} = 2$

$2 \times 8 = \boxed{}$

$16 \div \boxed{} = 8$

➡ *Pupil Book page 58*

Tables ×2, ×5, ×10

1 Colour the rest of this pattern to show the 2 times table. Write the answers at the end of each row.

2 × 1	2
2 × 2	4
2 × 3	6
2 × 4	
2 × 5	
2 × 6	
2 × 7	
2 × 8	
2 × 9	
2 × 10	
2 × 11	
2 × 12	

2 Use the 100 chart below.

a Circle all the multiples of 5.

b Underline all the multiples of 10.

c Colour all the blocks with even numbers yellow.

1	2	3	4	5	6	7	8	9	10
11	12	13	14	15	16	17	18	19	20
21	22	23	24	25	26	27	28	29	30
31	32	33	34	35	36	37	38	39	40
41	42	43	44	45	46	47	48	49	50
51	52	53	54	55	56	57	58	59	60
61	62	63	64	65	66	67	68	69	70
71	72	73	74	75	76	77	78	79	80
81	82	83	84	85	86	87	88	89	90
91	92	93	94	95	96	97	98	99	100

d What do you notice about the patterns?

➡ *Pupil Book page 59*

The ×3 table

1 Complete the table.

×	1	2	3	4	5	6	7	8				12
3	3	6							27	30		

2 Complete the table.

Number of counters	33	3	12	18	27	30	15	36	24	21		
How many groups of 3?	11										2	3

3 In each box, draw lines to match the numbers to the correct answers.

× 3	
5	33
9	3
1	18
11	15
7	21
6	27
2	9
8	6
3	24

÷ 3	
12	6
18	4
21	10
30	5
15	7
9	8
33	3
27	9
24	11

4 Ahmed waters his garden every three days during February. He starts on the 3rd of February.

Circle the days when he waters his garden.

How many times does he water the garden during the month?

February						
Mon	Tue	Wed	Thu	Fri	Sat	Sun
		1	2	3	4	5
6	7	8	9	10	11	12
13	14	15	16	17	18	19
20	21	22	23	24	25	26
27	28					

➡ *Pupil Book page 60*

Make equal groups

I have 18 counters.

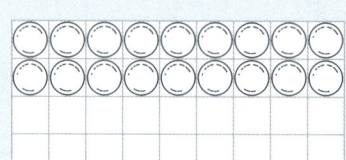

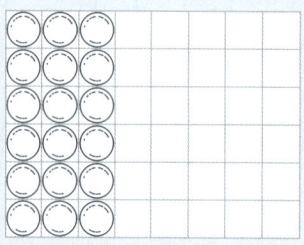

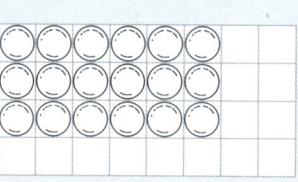

I can make 2 rows of 9.
$18 \div 2 = 9$

I can make 6 rows of 3.
$18 \div 6 = 3$

I can make 3 rows of 6.
$18 \div 3 = 6$

1 Draw four different ways of putting 12 counters into equal rows. Write a division fact for each array.

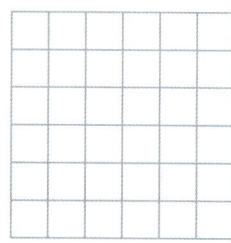

_____   _____   _____   _____

2 How many ways can you put 24 counters into equal rows?

_____

Draw all the ways you can find. Write a division fact for each array.

➡ *Pupil Book page 63*

Bar models

Look at this bar model.

100							
50				50			
25		25		25		25	

1 The top row in the bar model is equal to 100. The second row is divided into two sections. Each section is equal to 50.

Fill in the gaps.

a 50 + ☐ = 100

b 2 × ☐ = 100

c 100 ÷ 2 = ☐

d Half of ☐ = 50

2 Look at the third row. It is divided into four sections. Use the bar model to help you complete the number sentences.

a 25 + 25 = ☐

b 2 × 25 = ☐

c 4 × ☐ = 100

d $25 = \frac{1}{2}$ of ☐

e $25 = \frac{1}{4}$ of ☐

3 Look at the next three rows in the bar model. What numbers go in the sections in each row? Tell a partner how you worked it out.

fourth row: _____

fifth row: _____

sixth row: _____

➡ *Pupil Book page 64*

Multiples

1 Add up the number of sides to help you count on to the next multiple. Complete the sequences.

a △3 △6 △9 △ △ △ △ △ △ △

b ☐4 ☐8 ☐12 ☐ ☐ ☐ ☐ ☐ ☐ ☐

c ⬠5 ⬠10 ⬠ ⬠20 ⬠ ⬠ ⬠ ⬠ ⬠ ⬠

d ⬡6 ⬡12 ⬡ ⬡24 ⬡ ⬡ ⬡ ⬡ ⬡ ⬡

e ⯃8 ⯃16 ⯃ ⯃ ⯃ ⯃ ⯃ ⯃ ⯃ ⯃

2 Write:

a three multiples of 4 that are also multiples of 3 ☐ ☐ ☐

b two multiples of 4 that are also multiples of 6 ☐ ☐

c four multiples of 4 that are also multiples of 8. ☐ ☐ ☐ ☐

3 Write or say what patterns you notice.

_____

_____

_____

_____

_____

_____

_____

_____

_____

➡ *Pupil Book page 65*

Find the multiples

1 Circle the multiples of 2.

18	26	31	44	87	98	121
130	144	188	298	326	225	440
613	618	702	888	768	900	999

2 Circle the multiples of 5.

28	30	45	60	65	120	132
144	245	502	253	400	405	420
500	509	551	590	595	670	658

3 Circle the multiples of 10.

70	75	80	88	90	120	132
150	180	200	235	308	400	390
500	510	505	529	550	600	900

4 Circle the numbers that are multiples of both 2 and 5.

18	25	20	22	90	60	88
105	120	188	145	160	440	230
312	315	325	330	480	486	485

5 Which multiples have been used for each number line?

a
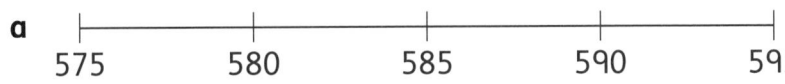
575 580 585 590 595

_____

b
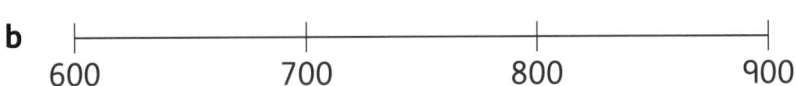
600 700 800 900

_____

c
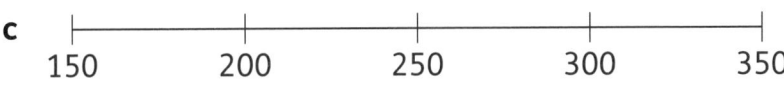
150 200 250 300 350

_____

➡ *Pupil Book page 65*

Division with some left over

1 Share the items among the children.

Are there any left over? Tick (✔) the correct box.

If there are some left over, write how many.

	Items to be shared	Number of children	Any left over?
a		2	Yes ☐ ☐ No ☐
b		4	Yes ☐ ☐ No ☐
c		3	Yes ☐ ☐ No ☐
d		4	Yes ☐ ☐ No ☐
e		7	Yes ☐ ☐ No ☐
f		4	Yes ☐ ☐ No ☐

▶ *Pupil Book page 66*

Multiplication facts

1 Use this grid to complete all the multiplication tables up to 10. You can use skip counting or multiples.

You can use counters if you want to check your answers.

For example, to check 2 × 3, put 2 counters in each of the first 3 rows.
Count up the total. 2 × 3 = 6

Write 6 in the block for 2 × 3. Then remove the counters.

You already know some of the multiplication facts. Only use counters when you need help.

×	1	2	3	4	5	6	7	8	9	10
1	1									
2	2	4	6				14			
3										30
4										
5					25					
6									54	
7			21				49		63	70
8								64	72	
9							63	72	81	90
10	10	20				60	70			100

➡ *Pupil Book page 66*

Multiplication patterns

1 Look for a pattern. Fill in the missing numbers.

a   20   18   16   ☐   ☐   ☐   ☐

b   5   10   ☐   ☐   25   ☐   ☐

c   10   20   30   ☐   ☐   ☐   ☐

d   3   6   9   ☐   ☐   ☐   ☐

e   12   ☐   16   18   ☐   ☐   ☐

f   4   8   12   ☐   ☐   ☐   ☐

g   36   32   ☐   ☐   ☐   ☐   ☐

2 Complete this multiplication grid.

×	1	2	3	4	5	6	7	8	9	10	11	12
2	2											
3		6										
4			12									
5				20								
10					50							

3 Double each number.

1   ☐   10   ☐   15   ☐

2   ☐   20   ☐   25   ☐

3   ☐   30   ☐   35   ☐

4   ☐   40   ☐   45   ☐

5   ☐   50   ☐

➡ *Pupil Book page 67*

Work with tens

1 Fill in the missing numbers in each table.

a

Number	7	4		2		8	
× 10		40	90		60		30

b

Number	12	15	16	18	21	23	25
× 10	120						

c

Number	20						
× 10	200	220	260	300	310	360	390

d

Number	40			47	48		
× 10		430	460			500	540

e

Number	68			88		95	
× 10		740	810		900		990

2 Fill in the missing numbers in each table.

a

Number	100	120	150	190	250	390	350	380	400
÷ 10	10			19					

b

Number	1000	970	940	870	600	340	800	810	700
÷ 10		97			60				

c

Number	200		50		140		760		150
÷ 10		82		11		27		4	

d

Number	100	1000	200	500	900	700	600	400	300
÷ 100				5					

➡ *Pupil Book page 67*

Doubling and halving

1 Complete the function machines. Fill in the missing numbers.

a
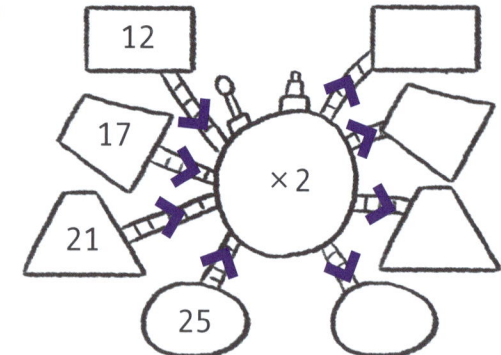

b
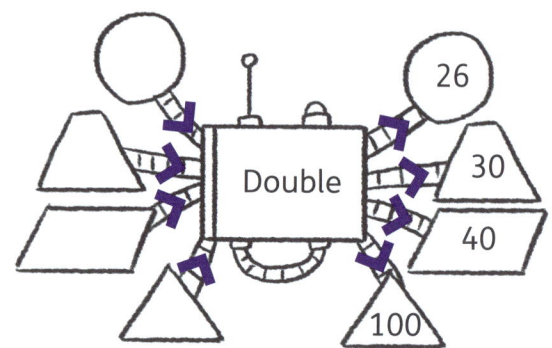

c
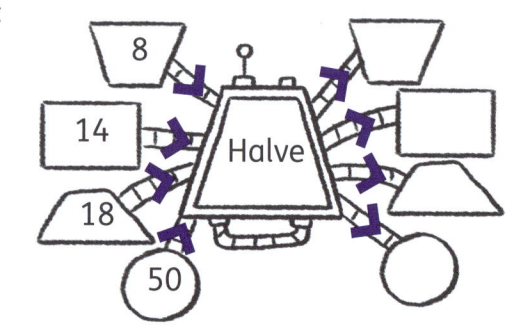

d
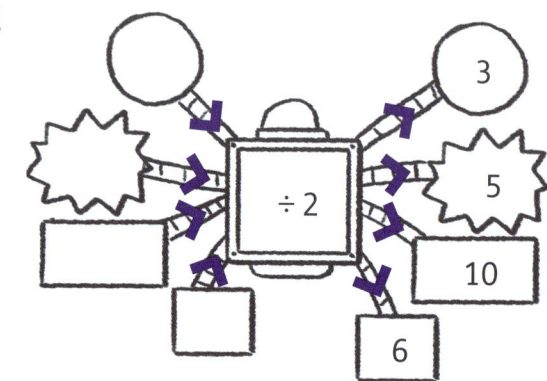

2 Fill in the missing numbers.

a

| 15 | $\xrightarrow{\text{double}}$ | | $\xrightarrow{\text{double}}$ | | $\xrightarrow{\text{double}}$ | |

b

| 36 | $\xrightarrow{\text{halve}}$ | | $\xrightarrow{\text{halve}}$ | |

c

| 24 | $\xrightarrow{\text{halve}}$ | | $\xrightarrow{\text{halve}}$ | | $\xrightarrow{\text{halve}}$ | |

3 Complete each bar model by writing the missing number.

a

100	
50	

b

200	
100	

c

600	
	300

d

800	
	400

➡ *Pupil Book page 68*

Perimeter and area

Measure perimeter

To find the perimeter, we measure and add up the lengths of all the sides.

1. Measure the sides of each shape to the nearest centimetre. Label the sides. Then add up the side lengths to find the perimeter. Show your working.

a

Perimeter:

b

Perimeter:

c

Perimeter:

d

Perimeter:

e

Perimeter:

f

Perimeter:

➡ *Pupil Book page 69*

Calculate perimeter

We can also use given measurements instead of measuring with a ruler.

1 Work out the perimeter in each question.

a

This strip of cardboard is 4 cm wide and 32 cm long. What is the perimeter of the strip of cardboard?

☐ + ☐ + ☐ + ☐ = ☐

Perimeter = ☐ cm

b

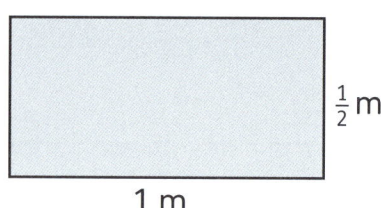

$\frac{1}{2}$ m

1 m

This rectangle is $\frac{1}{2}$ m wide and 1 m long. What is the perimeter of the rectangle?

☐ + ☐ + ☐ + ☐ = ☐

Perimeter = ☐ m

c

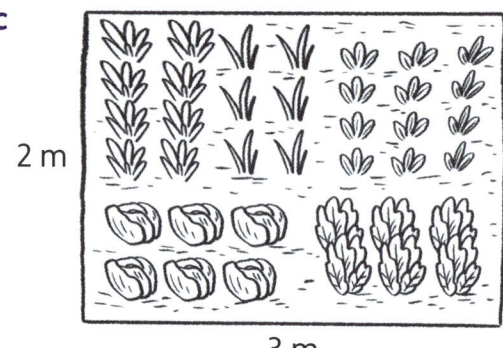

2 m

3 m

Work out the perimeter of this vegetable patch.

☐ + ☐ + ☐ + ☐ = ☐

Perimeter = ☐ m

d

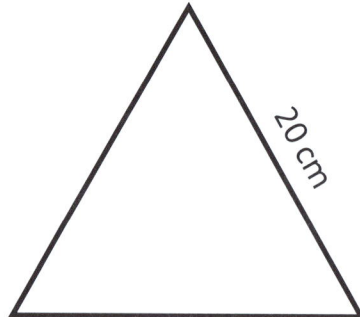

20 cm

Each side of this equilateral triangle is 20 cm long. What is the perimeter?

☐ × 3 = ☐

Perimeter = ☐ cm

2 What is the perimeter of a rectangle that has a length of 10 cm and a width of 5 cm?

☐

➡ *Pupil Book page 69*

Area

1 Count the squares to work out the area of each shape.
Each square on the grid is 1 cm².

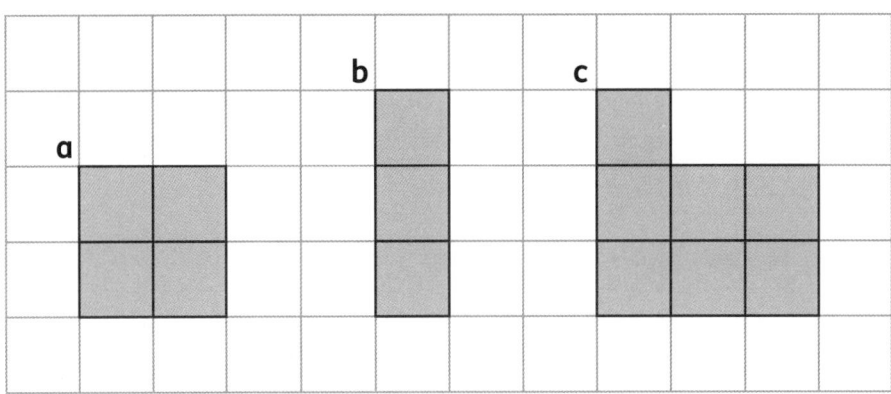

Area = ☐ cm²   Area = ☐ cm²   Area = ☐ cm²

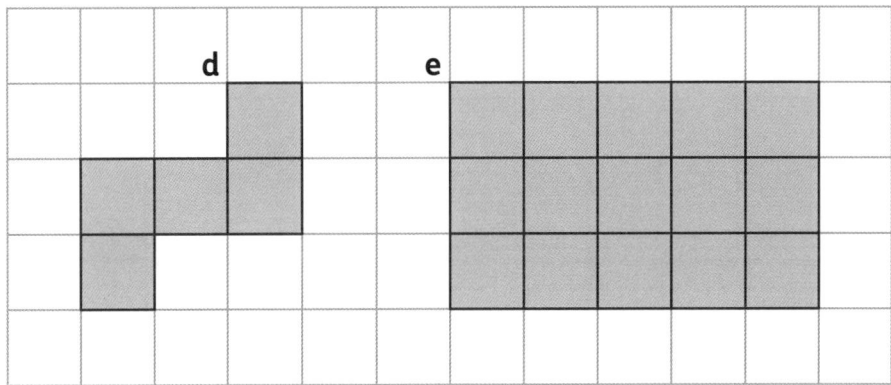

Area = ☐ cm²   Area = ☐ cm²

2 Draw as many different shapes as you can that have an area of 12 cm².

➡ *Pupil Book page 71*

Draw and measure

1 Draw rectangles and other polygons on the squared paper. Each square is 1 cm long and 1 cm wide. Find the perimeter and area of each shape. Write the perimeter and area next to each shape.

▶ *Pupil Book page 71*

Data

Make 20

1 **a** Cross out pairs of numbers that make 20. For example: 19 + 1.

Each time you cross out a pair, draw a tally mark in the table.

Add up your tallies to find out how many twenties there are in the box.

Remember, tally marks look like this:

| = 1   ⊞ = 5

6	16	7	X	13	4	19	15
11	17	5	18	2	17	9	16
5	7	3	12	5	11	15	6
12	4	15	17	5	1	13	18
3	5	19	16	14	14	3	2
8	3	17	15	8	4	15	9

Tally	Number of twenties

b What is the total of all the numbers in the box?

All the numbers in the box add up to: ☐

2 Work with a partner. Write ten different additions with a total of 19. Write two subtraction facts that you can work out from each addition. One is done for you.

Additions that make 19	Subtraction facts	
15 + 4 = 19	19 − 4 = 15	19 − 15 = 4

➡ *Pupil Book page 72*

Use symbols

Here are four symbols from pictograms.

= 1 person	= 2 ice creams	= 1 car	= 2 fish

1 Write the amounts that these symbols show.

a	**b**	**c**	**d**
e	**f**	**g**	**h**

2 Draw the symbols to show these amounts.

a 3 people	**b** 7 ice creams
c 3 cars	**d** 5 fish

➡ Pupil Book page 74

Surveys

1 Do a survey to find out which ice cream flavours your classmates like best.

Flavour	Tally	Total
Vanilla		
Strawberry		
Chocolate		
Lime		
Mango		
Coconut		

2 Draw a pictogram to show the results of your survey. Use your own key.

Our favourite ice cream flavours

Key: [　　　] = [　] children

➡ *Pupil Book page 74*

Statistics **57**

Sort data

1 Look at the picture. Answer the questions.

a How many children are there? ☐

b Complete this tally table about what the children are wearing.

	Tally
Wearing shorts	
Wearing a skirt	
Wearing long trousers	

c Complete this tally table.

	Tally
Wearing glasses	
Not wearing glasses	

d Complete this Carroll diagram about what the children are wearing. Write the number of children in each section of the table.

	Wearing a skirt	Not wearing a skirt
Wearing glasses		
Not wearing glasses		

2 Collect information about the pupils in your class.

	Tally
Wears glasses	
Does not wear glasses	
Likes dancing	
Does not like dancing	
Has short hair	
Does not have short hair	

➡ *Pupil Book page 75*

Bar charts

Five children ran 1500 metres. Here are their times:

Sarah: 8 minutes

Ayiz: 12 minutes

Dan: 14 minutes

Mishka: 11 minutes

Anna: 13 minutes

1 Use the information to complete the bar chart.

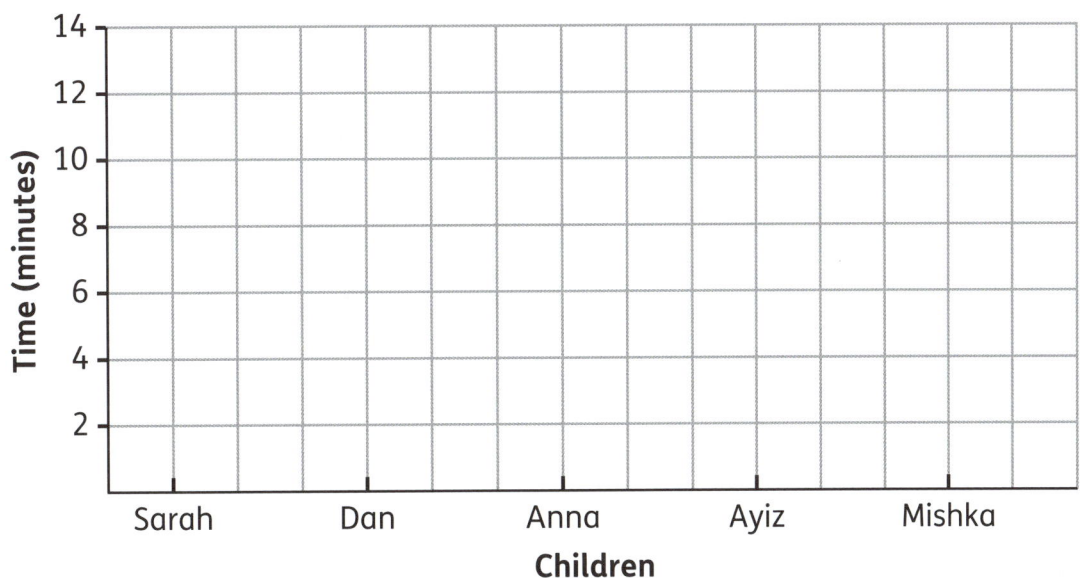

2 Answer these questions about the bar chart.

a Write a title for the bar chart. _____

b Who had the fastest time? _____

c Who had the third fastest time? _____

d How much longer did Dan take than Sarah? _____

3 Write a question about the bar chart for a partner to answer.

_____

➡ *Pupil Book page 75*

Draw bar charts

1 Find out which sandwich fillings your classmates like.

Make a tally table.

Filling	Tally	Total
Cheese		
Salad		
Avocado		
Jam		
Chocolate		

2 Draw a bar chart to show your results.

Our favourite sandwich fillings

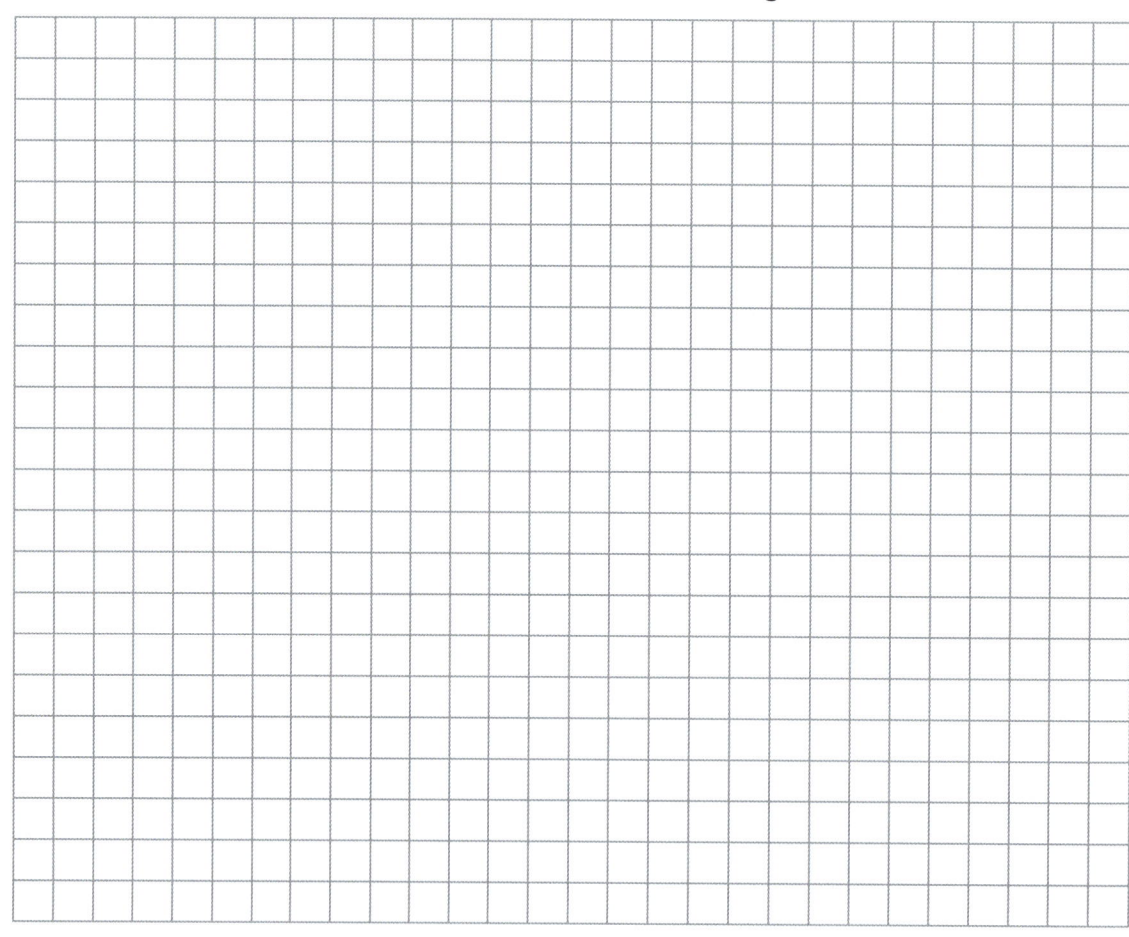

Number of pupils

Fillings

➡ *Pupil Book page 75*

Sports survey

This table shows the sports that the pupils in Year 3 like best.

Sport	Number of pupils
Football	14
Basketball	18
Swimming	20
Tennis	8
Athletics	16

1 Complete the bar chart to show this information.

Remember to write a title for the chart.

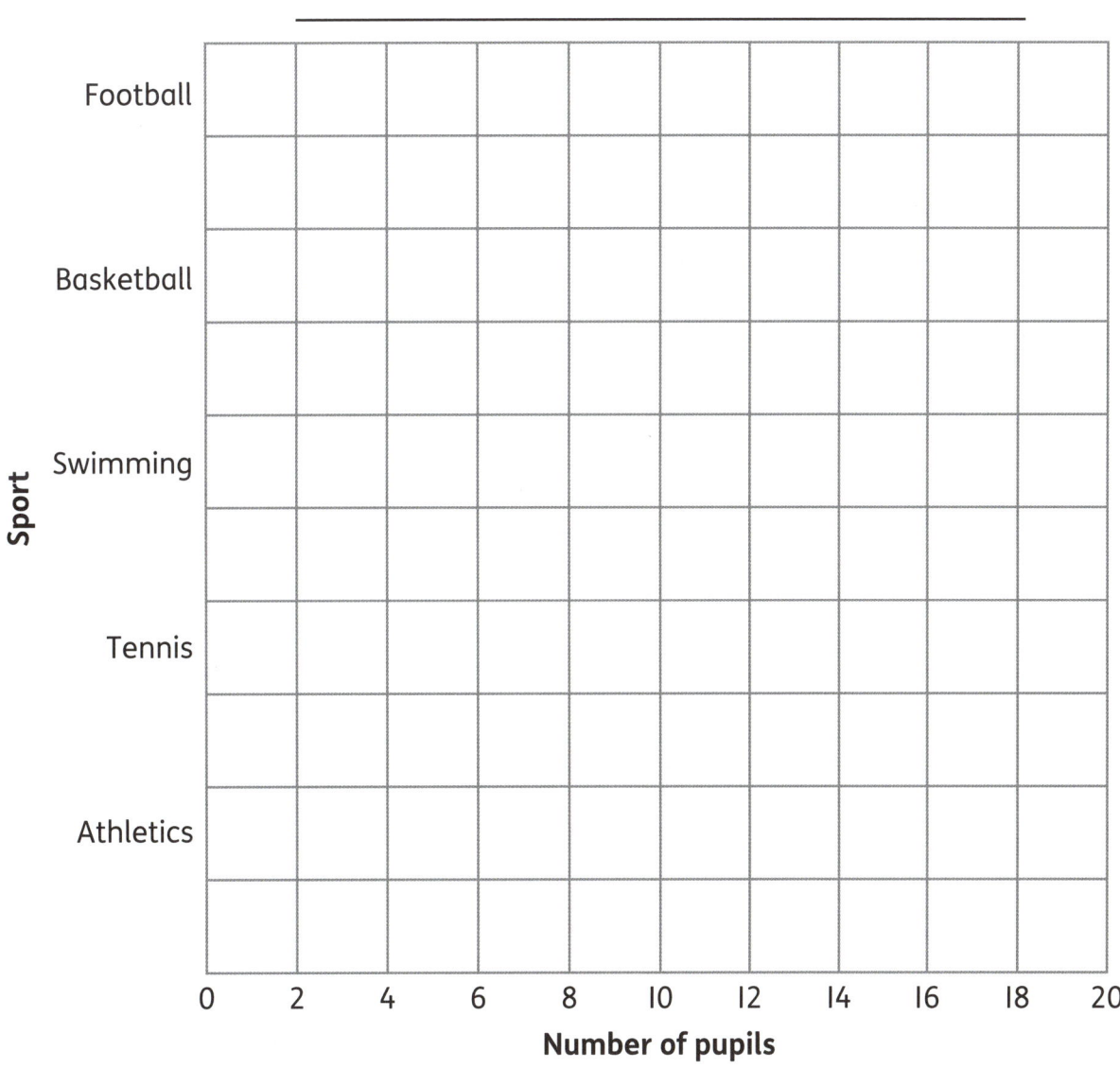

2 Which sport is the most popular? _____

3 Which sport is the least popular? _____

➡ *Pupil Book page 75*

Venn diagrams

1 Write each number in the correct place in the Venn diagram. Cross off each number when you write it.

> You can put numbers outside the circles if they do not fit into either set.

| 22 | <u>29</u> | 18 | <u>35</u> | 63 | 45 | 27 | 51 |
| 72 | <u>48</u> | 20 | <u>24</u> | 32 | <u>15</u> | 21 | <u>12</u> |

All numbers

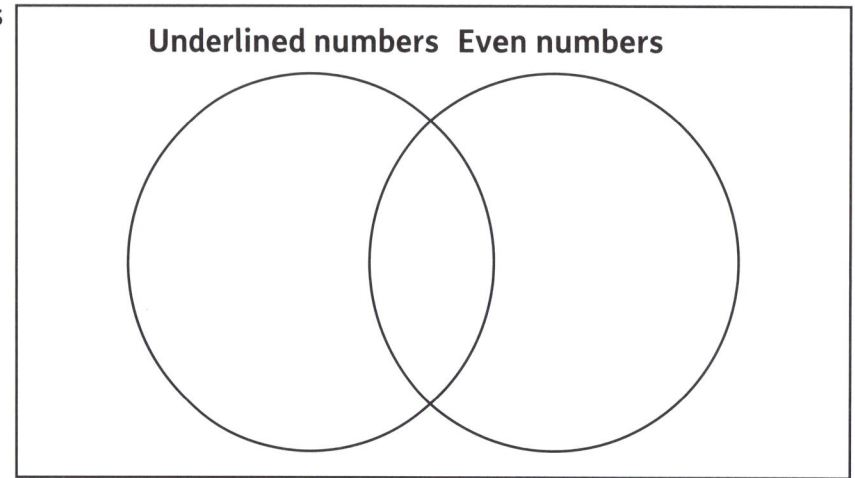

2 Colour each item the correct colour. Then sort them into the Venn diagram.

banana   chocolate   sun   cheese

broccoli   orange   lemon   sunflower

All things

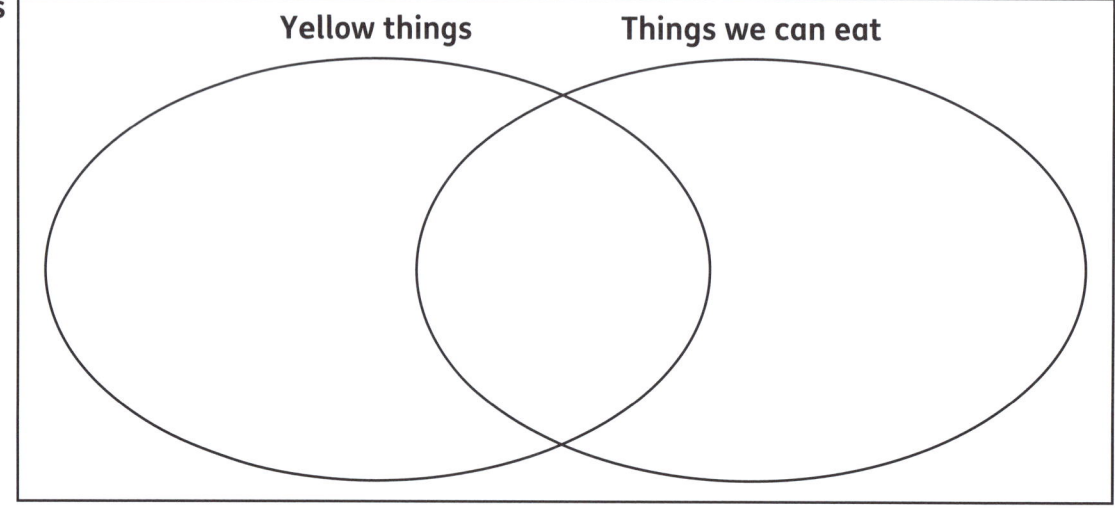

➡ *Pupil Book page 76*

Carroll diagrams

1 Draw the shapes in the correct boxes in this Carroll diagram.

	Grey	Not grey
Has right angles		
Has no right angles		

2 Sarita asked her friends whether they like mangoes and bananas. This is her Carroll diagram.

	Likes mangoes	Doesn't like mangoes
Likes bananas	5	3
Doesn't like bananas	1	2

a How many friends did Sarita ask? ☐

b How many friends like mangoes and bananas? ☐

c How many friends don't like mangoes or bananas? ☐

d How many friends like mangoes, but not bananas? ☐

e How many friends like bananas, but not mangoes? ☐

3 Write the numbers in the correct boxes in this Carroll diagram.

	Even	Not even
Multiple of 5		
Not a multiple of 5		

1	2	3	4	5
6	7	8	9	10
12	15	20	26	
30	35	50		

➡ *Pupil Book page 77*

A mini-beast investigation

Mini-beasts are tiny animals such as worms, termites and spiders.

1 Look at the pictures carefully. In which place would you **not** find many mini-beasts? Why not?

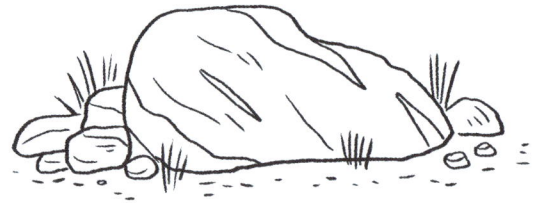

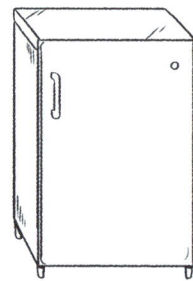

_____

_____

_____

2 Work in groups to count mini-beasts in different places. Your teacher will give you instructions.

Choose four different places. Count how many mini-beasts you find.

Record your data in this table.

Place	Number of mini-beasts

3 Use your table to answer these questions.

a Where did you find the most mini-beasts? _____

b Where did you find the fewest mini-beasts? _____

c How many mini-beasts did you find altogether? [ ]

➡ *Pupil Book page 77*

3D shapes

Faces, edges and vertices

1 Write the name of each 3D shape. You can use the names in the box to help you. Then write the number of faces, edges and vertices.

cylinder	triangular prism	cuboid	cube	pyramid

Name	Number of faces	Number of edges	Number of vertices
_____			
_____			
_____			
_____			
_____			

2 Each set of dots makes a picture of a 3D shape. Can you find a way to join the dots and make each shape?

> Both the 3D shapes are in the table above.

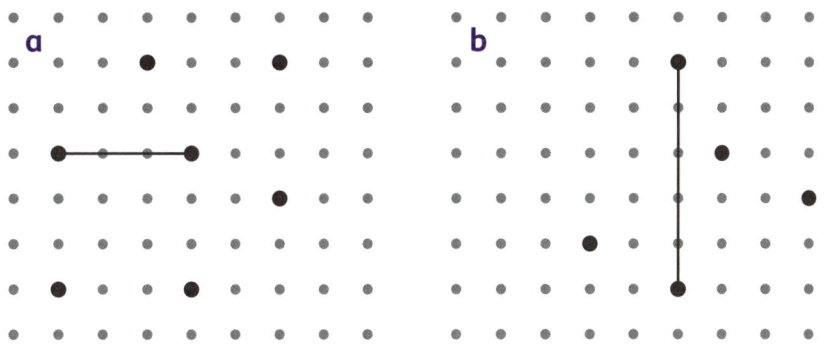

a   b

➡ *Pupil Book page 81*

Explore 3D shapes

1 Samantha used these 3D shapes to make a model.

Colour each shape a different colour. Write the name of each shape.

_____   _____   _____

_____   _____   _____

2 Look at this picture of Samantha's model. Colour the shapes in the picture to match your coloured shapes.

3 Look at Samantha's model. Count how many there are of each shape. Complete this table.

Shape	Number used
Cylinder	
Sphere	
Pyramid	
Triangular prism	
Cone	
Cuboid	

➡ *Pupil Book page 82*

Draw 3D shapes

1 Look again at Samantha's model on page 66.

a What did Samantha use the pyramid for? _____

b What shapes are the houses made from? _____

c Which shape did Samantha use most often? _____

d Which shape did Samantha use least often? _____

e Samantha used four cylinders for lamp posts.

What else did she use cylinders for? _____

2 Draw your own buildings and other structures below. Label the shapes you can see in your drawings.

➡ *Pupil Book page 82*

Build your own cube

1 Trace or copy this net onto thin card and cut it out.

Fold the net along the dotted lines to make your cube.

Use the tabs to glue the faces together.

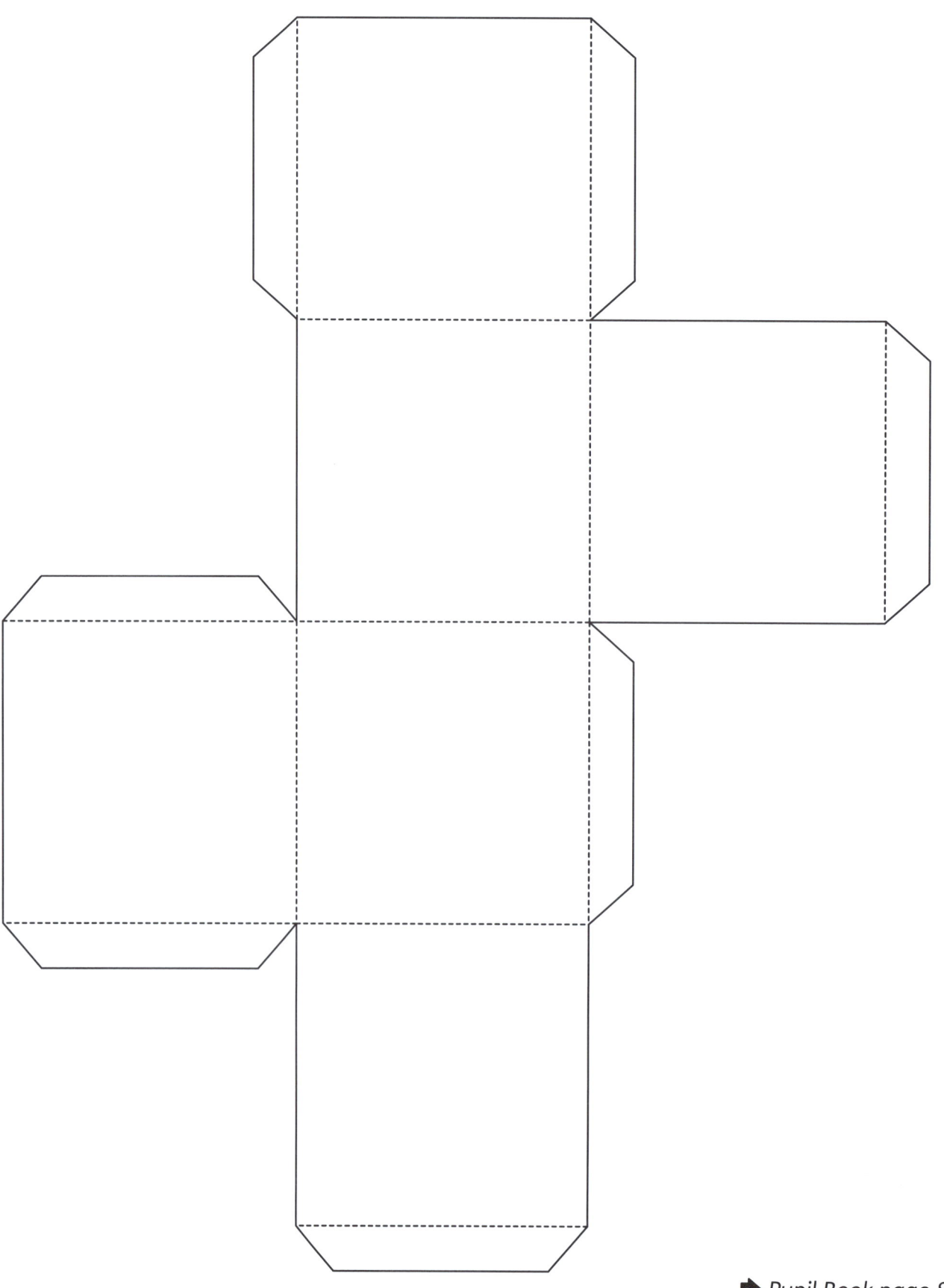

➡ *Pupil Book page 84*

Position, direction and movement

Clockwise and anti-clockwise

1 Imagine you are at X. Complete the table. The first row is done for you.

Starting position	Turn	End position
Facing the tree	$\frac{1}{2}$ turn clockwise	Facing the swing
	$\frac{1}{2}$ turn anti-clockwise	
Facing the swing	$\frac{1}{4}$ turn anti-clockwise	
	$\frac{1}{4}$ turn clockwise	
Facing the pond	$\frac{3}{4}$ turn anti-clockwise	
	$\frac{3}{4}$ turn clockwise	
Facing the house	$\frac{1}{2}$ turn clockwise	
	$\frac{3}{4}$ turn clockwise	
Facing the tree	$\frac{1}{4}$ turn anti-clockwise	
	$\frac{3}{4}$ turn clockwise	

2 Draw a small map of a treasure island. Draw a tree, a boat, a cave and some treasure. Make up some starting positions and turns of your own. Ask a partner to work out the end positions.

➡ *Pupil Book page 86*

More turns

1 A snail took this route around some tree stumps.

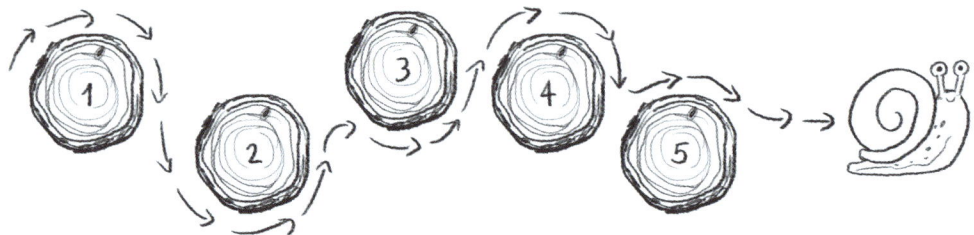

a Around which tree stumps did the snail travel clockwise?

_____

b Around which tree stumps did the snail travel anti-clockwise?

_____

2 Another snail took a different route. It went anti-clockwise around stumps 1, 2 and 4. It went clockwise around stumps 3 and 5.

Draw the snail's route using arrows.

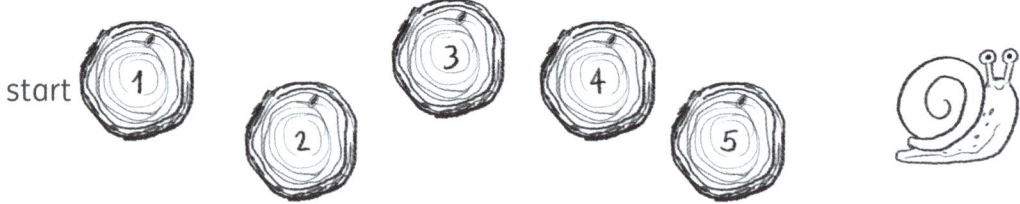

3 Draw some circles using an anti-clockwise motion. Draw some circles using a clockwise motion.

Which motion felt more comfortable? Suggest some reasons for your answer.

➡ *Pupil Book page 86*

Position

This is part of a seating plan for a class visit to the theatre.

Back

	A	B	C	D	E
4					Zayed
3					
2	Jo		Thandi	Cara	Luke
1			Malala		

Front

1. Write the names of these pupils in the correct places on the plan.

   a Dahlia sits in seat B4.

   b Rick sits two rows in front of Dahlia.

   c Sipho sits in E1.

   d Amani sits next to Sipho.

   e Bess sits in C3.

   f Nina sits behind Bess.

   g Ella sits in A1.

2. Write the positions of these pupils.

   a Zayed ☐

   b The person in front of Luke ☐

   c Cara ☐

   d Jo ☐

   e The person between Rick and Cara ☐

3. Some pupils move seats. Write their new positions.

   a Zayed moves one seat forwards. ☐

   b Jo moves one row back. ☐

   c Dahlia moves to the empty seat in the front row. ☐

➡ *Pupil Book page 87*

Position on a grid

1 Draw the shapes in the correct squares on the grid.

A4 ⬤   C5 ✶   D9 ◯   E1 ▱

F3 ⬡   D4 ▲   G4 ◻   I2 ⯃

J9 ▭   E8 △   H2 ◯   A10 ⚫

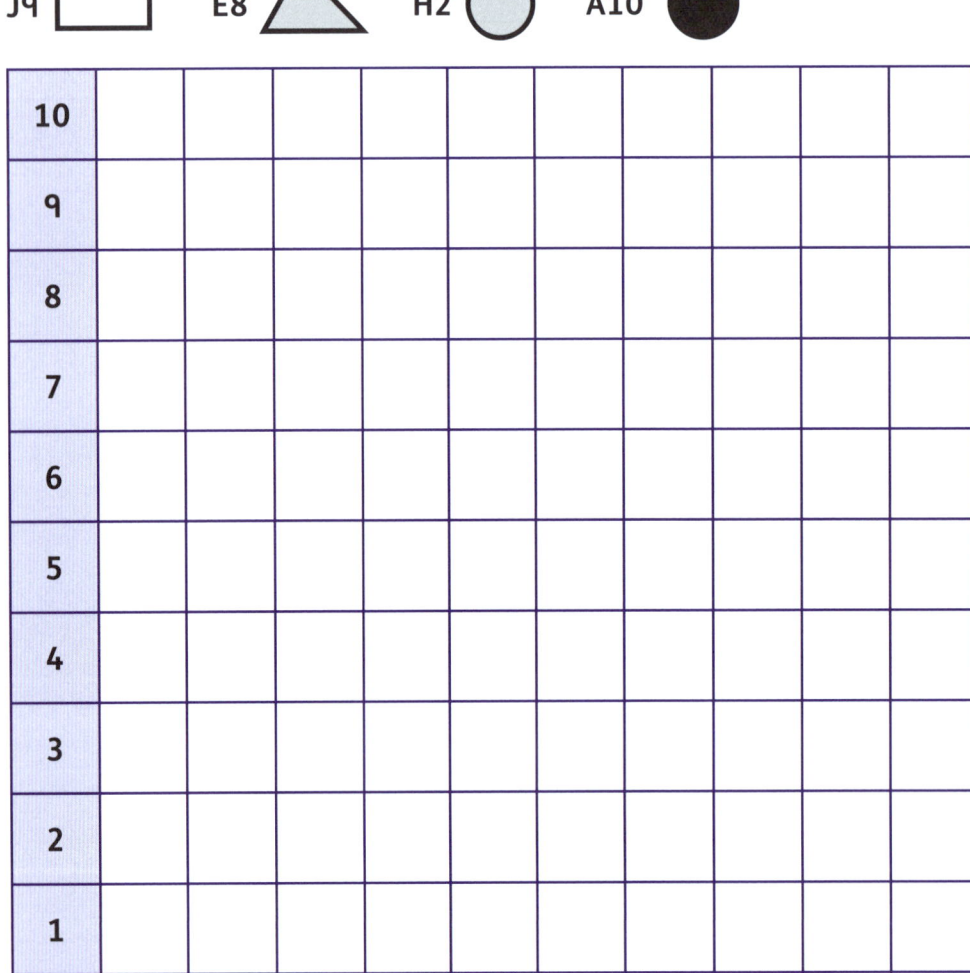

2 Write your own directions to move from:

a the hexagon to the black circle.

_____

b the rectangle to the star.

_____

➡ *Pupil Book page 87*

Fractions

1. Shade the correct fraction of each shape.

a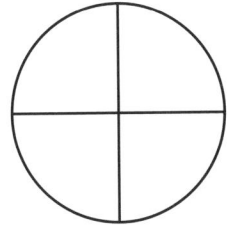

one-half

b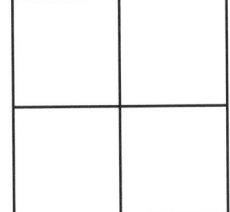

three-quarters

c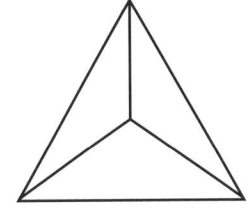

two-thirds

2. Complete the table.

Fraction in words	Fraction in numbers	Number of equal parts in the whole
one-quarter	$\frac{1}{4}$	4
one-half		
two-thirds		
three-fifths		
five-sixths		
four-tenths		

3. Colour the necklaces correctly.

a $\frac{1}{2}$ of the beads are green. The rest are yellow.

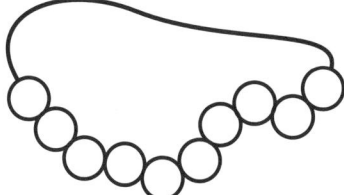

b $\frac{1}{4}$ of the beads are green. $\frac{1}{4}$ of the beads are yellow. The rest are purple.

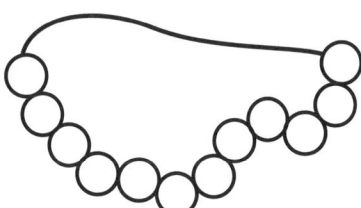

4. a Shade $\frac{1}{2}$ of the circles.

b Shade $\frac{1}{4}$ of the triangles.

➡ *Pupil Book page 90*

More fractions of shapes

1 Colour the fraction given. Write the fraction that is not shaded.

a

$\frac{1}{5}$

[ ] not shaded

b

$\frac{3}{4}$

[ ] not shaded

c

$\frac{1}{3}$

[ ] not shaded

d

$\frac{3}{10}$

[ ] not shaded

e

$\frac{1}{2}$

[ ] not shaded

f

$\frac{4}{8}$

[ ] not shaded

g

$\frac{2}{3}$

[ ] not shaded

h

$\frac{1}{8}$

[ ] not shaded

i

$\frac{1}{4}$

[ ] not shaded

2 Complete the fraction wall. Write the missing fractions.

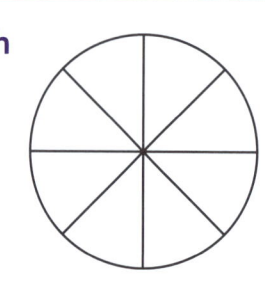
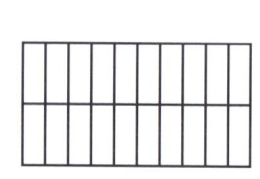

1 whole			
$\frac{1}{2}$		$\frac{1}{2}$	

➡ *Pupil Book page 91 and page 92*

Fractions of amounts

1 Count the squares in each chocolate bar. Complete the number sentences.

a

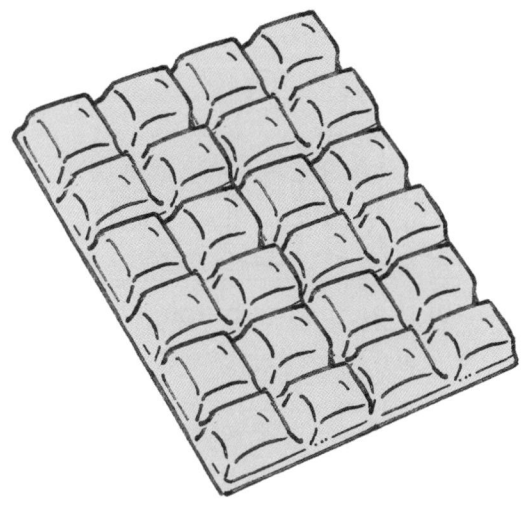

$\frac{1}{2}$ of the squares = [ ]

$\frac{1}{3}$ of the squares = [ ]

$\frac{1}{4}$ of the squares = [ ]

b

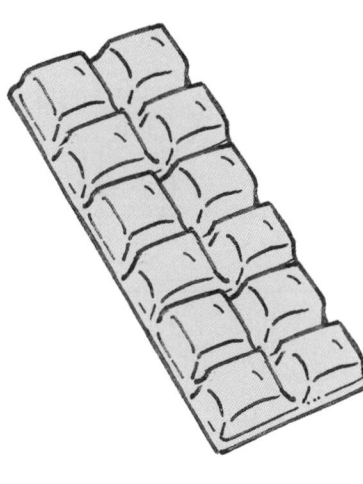

$\frac{1}{2}$ of the squares = [ ]

$\frac{1}{4}$ of the squares = [ ]

$\frac{3}{4}$ of the squares = [ ]

c

$\frac{1}{2}$ of the squares = [ ]

$\frac{1}{4}$ of the squares = [ ]

2 Write the missing number.

a $\frac{1}{3}$ of [ ] = 3

b $\frac{1}{2}$ of [ ] = 5

c $\frac{1}{4}$ of [ ] = 4

d $\frac{1}{2}$ of [ ] = 12

e $\frac{1}{3}$ of [ ] = 1

f $\frac{1}{4}$ of [ ] = 9

➡ *Pupil Book page 95*

Colour the fraction

1 Colour or draw on each picture to show the correct fractions.

a $\frac{1}{2}$ of the bottles are full.

b $\frac{1}{3}$ of the ladybirds have no spots.

c $\frac{1}{4}$ of the cars are purple.

d $\frac{1}{5}$ of the fish are green.

e $\frac{1}{2}$ of the birds are yellow.

f $\frac{1}{6}$ of the beads are striped.

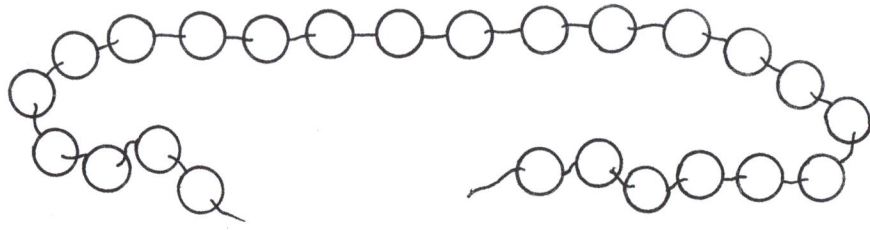

2 Choose some parts of each shape to colour. Write the fraction of each shape that you colour.

a

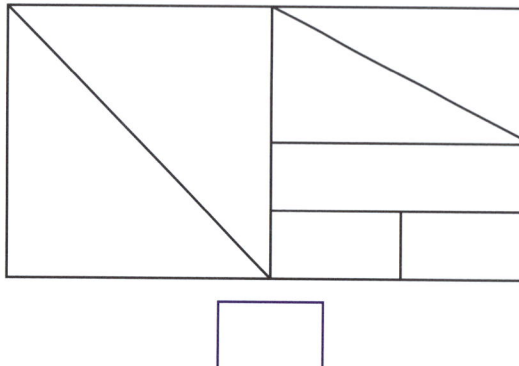

b

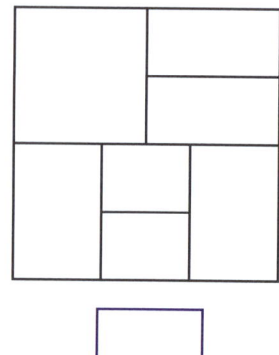

➡ *Pupil Book page 95*

Add and subtract fractions

 $\frac{3}{8}$ was shaded.

Add $\frac{2}{8}$ more.

$\frac{3}{8} + \frac{2}{8} = \frac{5}{8}$

1 Shade more parts of each shape to show the addition. Then write the missing answers.

a

$\frac{1}{4} + \frac{2}{4} = \boxed{}$

b

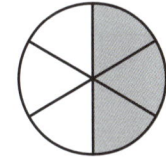

$\frac{3}{6} + \frac{2}{6} = \boxed{}$

c

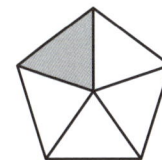

$\frac{1}{5} + \boxed{} = \frac{4}{5}$

2 Calculate.

a $\frac{1}{3} + \frac{1}{3} = \boxed{}$

b $\frac{2}{5} + \frac{1}{5} = \boxed{}$

c $\frac{9}{10} - \frac{3}{10} = \boxed{}$

d $\frac{3}{8} - \frac{2}{8} = \boxed{}$

e $\frac{4}{10} + \frac{3}{10} = \boxed{}$

f $\frac{1}{2} + \frac{1}{2} - \frac{1}{2} = \boxed{}$

3 Look at each calculation. Is it correct or incorrect? Explain your answer. You can use sketches to help you explain.

Calculation	Correct or incorrect?	My reasons
a $\frac{8}{9} + \frac{1}{9} = 1$		
b $\frac{4}{5} - \frac{1}{5} = \frac{1}{4}$		
c $\frac{7}{10} - \frac{7}{10} = 0$		
d Jean-Luc has $\frac{3}{4}$ of a metre of fabric. He uses $\frac{1}{2}$ m. He has $\frac{1}{2}$ m left.		

4 Marielle has $\frac{9}{10}$ of a metre of wood. She cuts off $\frac{2}{10}$ of a metre.

How much is left? _____

➡ *Pupil Book page 100*

Capacity and temperature

Measure capacity

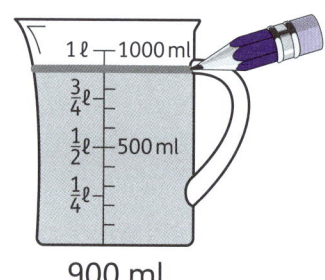

900 ml

1 Draw the liquid in the jugs to show each amount.
The one opposite is done for you.

a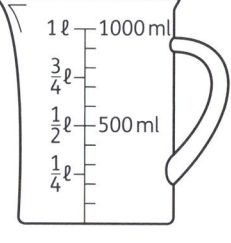

500 ml

b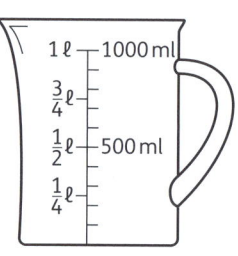

700 ml

c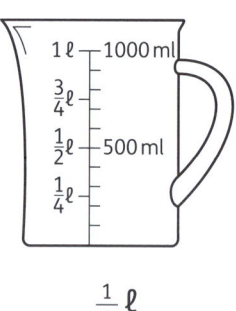

$\frac{1}{4}$ ℓ

d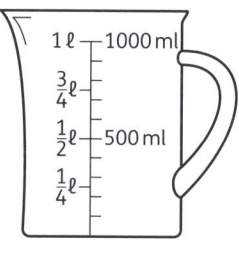

1 ℓ

e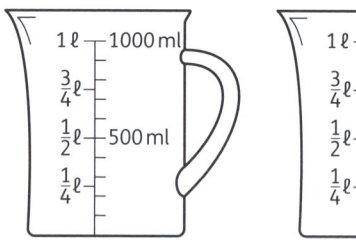

1 ℓ 200 ml

f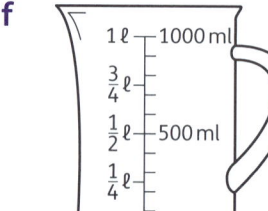

1 ℓ 700 ml

g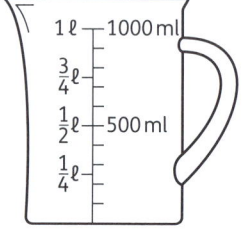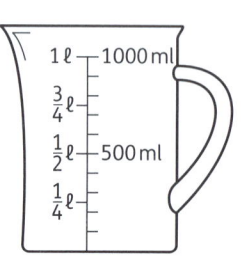

$1\frac{3}{4}$ ℓ

h

1 ℓ 900 ml

➡ *Pupil Book page 101*

Work with capacity

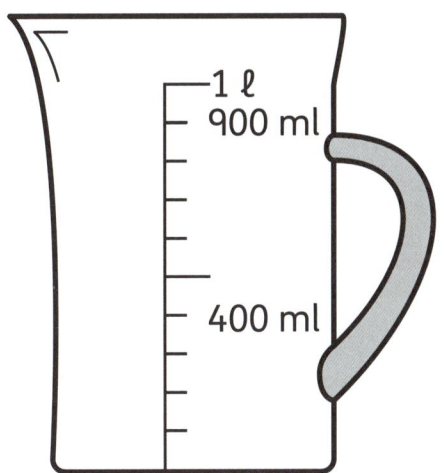

- 1 ℓ
- 900 ml
- 400 ml

1 Some labels on the scale of this jug are missing.
Write the missing labels on the jug.

2 Complete the missing capacities.

a 1000 ml = ☐ ℓ

b $\frac{1}{2}$ litre = ☐ ml

c $\frac{1}{4}$ litre = ☐ ml

3 Maya has some containers with different capacities.

D

B

C

A

50 ml

100 ml

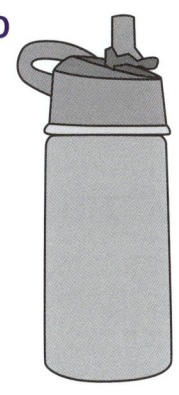

250 ml

750 ml

She fills container C four times to make 1 litre. Describe three other ways she can use the containers to make 1 litre.

_____

_____

_____

_____

4 About how many millilitres or litres are there in each container? Circle the best estimate.

a

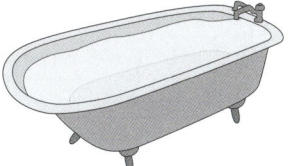

800 ml

or

80 ℓ

b

5 ℓ

or

150 ml

c

HONEY

370 ℓ

or

370 ml

➡ *Pupil Book page 104*

Temperature

1 Colour the scale on each thermometer to show the correct temperature.

a The temperature inside a cloud is about 1 °C.

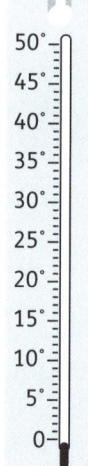

b The temperature of this seawater is about 22 °C.

c The air temperature at this beach is about 27 °C.

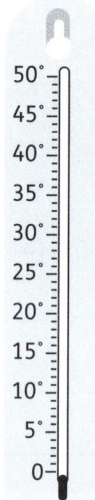

d Candle wax melts at about 37 °C.

e The temperature inside this car's engine is about 75 °C.

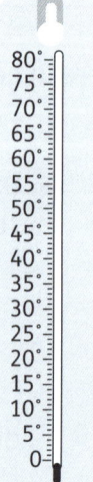

f A hot bath is about 40 °C.

➡ *Pupil Book page 105 and page 106*

Work with temperature

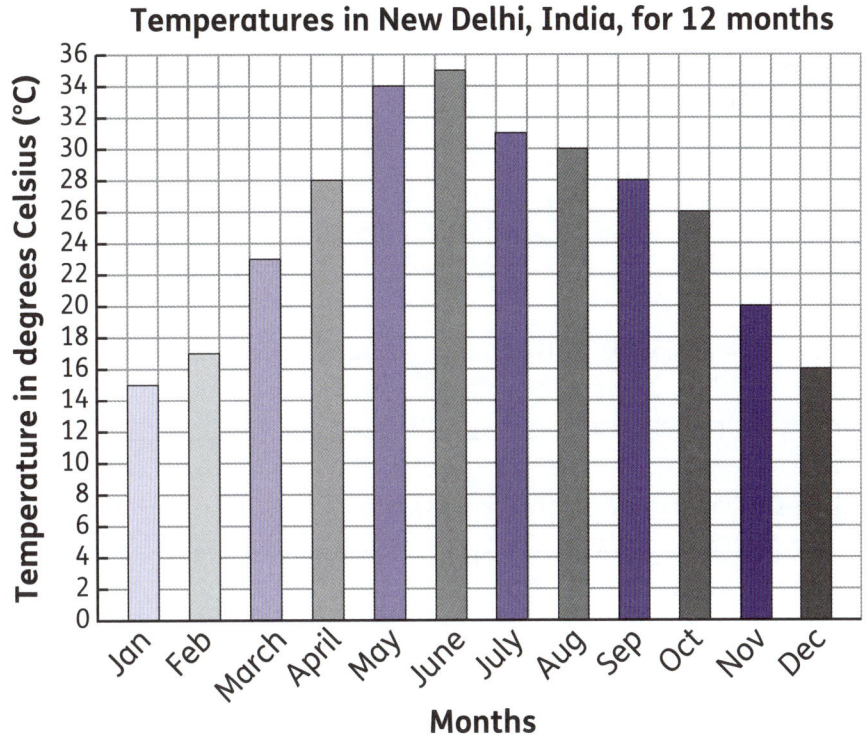

Temperatures in New Delhi, India, for 12 months

1 a What does the chart show? _____

   b Which are the three coolest months?

   _____   _____   _____

   c Which are the three hottest months?

   _____   _____   _____

2 In which months were the temperatures:

   a greater than 20 °C but less than 30 °C?

   _____   _____   _____   _____

   b greater than 30 °C?

   _____   _____   _____

3 Choose four of the months and write the temperature for each month.

   _____   _____   _____   _____

4 Imagine you are going to New Delhi in June. What kinds of clothing will you pack?
   Give reasons for your answer.

   _____

   _____

➡ *Pupil Book page 107*

Possible outcomes

1 Complete the statements.

a Today it is certain that _____.

b This afternoon it is possible that _____.

c This evening it is unlikely that _____.

2 Draw two pictures of events that are impossible.

3 Colour each spinner so that it has the possible outcomes given.

a

Three possible outcomes, all equally likely.

b

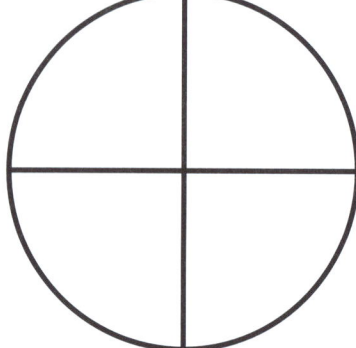

Two possible outcomes, one much more likely than the other.

➡ *Pupil Book page 109*

Probability experiments

For this experiment, you will need a bag of sweets with different colours.

1 **a** The type of sweets I chose: _____

Draw a picture of the packet.

b Mass of the sweets: ☐

c Number of sweets in the packet: ☐

d Tally table of colours:

🍬	🍬	🍬	🍬	🍬	🍬	🍬	🍬
🍬	🍬	🍬	🍬	🍬	🍬	🍬	🍬

e Colour that occurs most often in this packet: _____

Colour that occurs least often: _____

f Prediction: If I choose 10 sweets without looking, I think I will get these colours:

🍬 🍬 🍬 🍬 🍬 🍬 🍬 🍬 🍬 🍬

g Experiment: Put all the sweets back in the packet. Without looking, take 10 sweets out. Record the colours.

Actual outcome: 🍬 🍬 🍬 🍬 🍬 🍬 🍬 🍬 🍬 🍬

h How well did you predict the outcome? _____

_____

i Which colours occurred most and least often in your actual outcome?

_____

_____

_____

➡ *Pupil Book page 110*

Time

Show and tell the time

1 Draw or write the time on both watches.

The one opposite is done for you.

five minutes past eight

a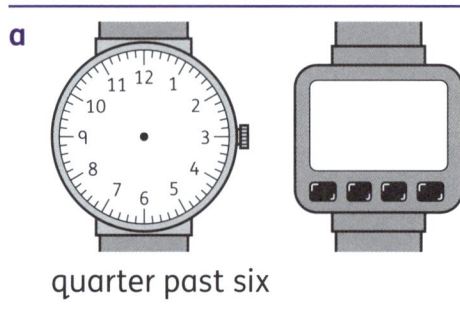

quarter past six

b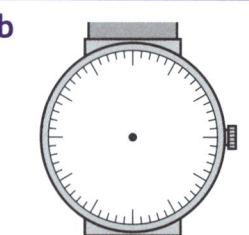

ten to seven

c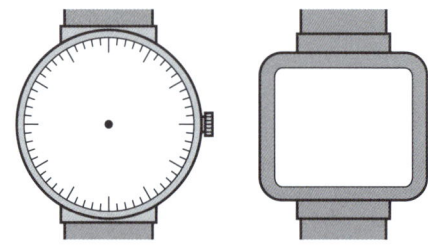

twenty-five past twelve

d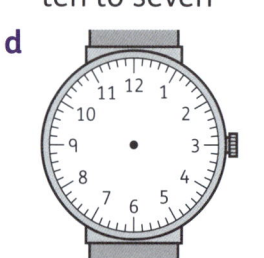

quarter to nine

2 Read the time on one watch in each pair.
Draw or write the same time on the other
watch. Then write the time in words.
The one opposite is done for you.

five minutes past five

a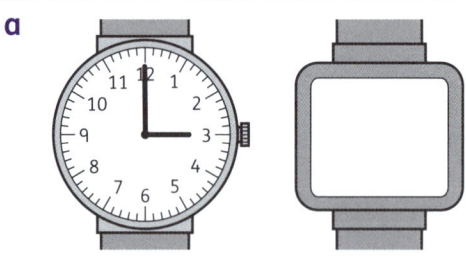

b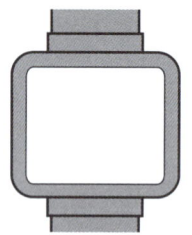

c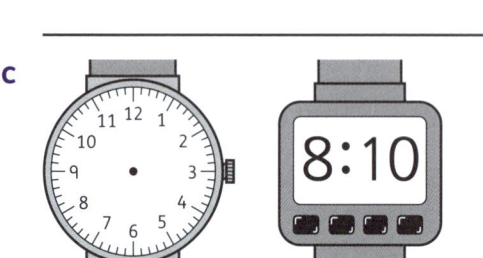

8:10

d

12:50

➡ *Pupil Book page 111 and page 112*

Timetables

> The 12-hour clock uses the numbers 1 to 12 for the hours. We write a.m. for morning times and p.m. for times after 12 noon.

1 Write these times using the 12-hour clock.

a   5 o'clock in the afternoon   _____

b   half past 3 in the morning   _____

c   quarter to 7 in the morning   _____

d   twenty to 4 in the afternoon   _____

2 Write these times in words.

a   6:35 p.m. _____   b   12:30 p.m. _____

c   11:15 a.m. _____

At an animal park, visitors take the bus to the animal viewing areas. They can get off at any area and get on again later.

Area	Bus A	Bus B	Bus C
Bus stop	9:30 a.m.	10:00 a.m.	10:30 a.m.
Tropical birds	9:35 a.m.	10:05 a.m.	10:35 a.m.
Giraffes	9:40 a.m.	10:10 a.m.	10:40 a.m.
Zebras	9:50 a.m.	10:20 a.m.	10:50 a.m.
Elephants	10:00 a.m.	10:30 a.m.	11:00 a.m.
Cheetahs	10:15 a.m.	10:45 a.m.	11:15 a.m.
Bus stop	10:40 a.m.	11:10 a.m.	11:40 a.m.

3 How long does it take to get from:

a   the bus stop to the tropical birds _____

b   the giraffes to the zebras _____

c   the elephants to the cheetahs _____

d   the cheetahs to the bus stop? _____

4 How often do the buses leave from the bus stop? Circle the correct answer.

every hour   every half hour   every ten minutes

5 Ann takes Bus A from the bus stop to the elephants.
She stays for 1 hour.

a   Which bus must she take to get back to the bus stop?   _____

b   What time will she get back to the bus stop?   _____

➡ *Pupil Book page 115 and page 116*

Estimate time

1 You will need a watch or a stopwatch for this activity. Work with a partner.

Look at the pictures. Estimate how long each activity will take. Colour the correct block in the table to record your estimate.

Do each activity. Time how long it takes you to the nearest minute. Write each time below and tick (✔) the box if your prediction was correct.

a

Put on your jacket.

_____ ☐

b

Bounce a ball 20 times.

_____ ☐

c

Walk around the playground 5 times.

_____ ☐

d

Count backwards from 100 to 0.

_____ ☐

e

Write all the numbers from 600 to 800.

_____ ☐

f

Copy 50 words from a book.

_____ ☐

	Activity	Less than 1 minute	1 to 5 minutes	5 to 15 minutes	More than 15 minutes
a	Put on your jacket				
b	Bounce a ball 20 times				
c	Walk around the playground 5 times				
d	Count backwards from 100 to 0				
e	Write the numbers 600 to 800				
f	Copy 50 words from a book				

➡ *Pupil Book page 117 and page 118*

Time in minutes

Six pupils timed how long it takes them to get to school each day.

Sunan   30 minutes   Ayesha   45 minutes

Taha   35 minutes   Kehinde   40 minutes

Nilar   25 minutes   Rubina   50 minutes

1 Draw a pointer on each timer to show how long each pupil took.

Sunan

Ayesha

Taha

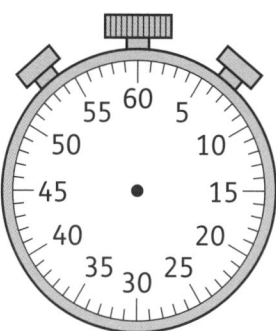

Kehinde

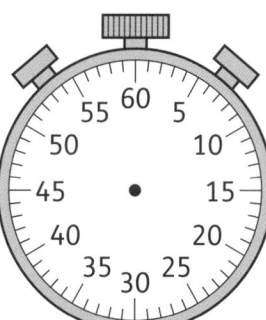

Nilar

Rubina

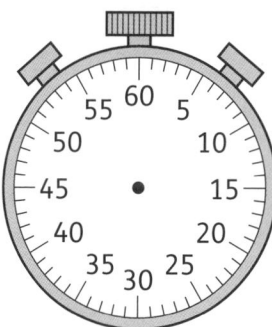

2 Write the times in order from longest to shortest.

_____   _____   _____

_____   _____   _____

3 What is the difference between the longest time and the shortest time?

_____

4 Who takes the longest to get to school? _____

5 Who gets to school in the shortest time? _____

➡ *Pupil Book page 119 and page 120*

Calendars and dates

1 Fill in the missing information on each calendar.

February						
Mon	Tues					
			2	3	4	
	7	8				
13						

September					
Sun	1	8			
Mon	2				
	3				
	4				
	5				
	6				
	7				

2 Write the correct dates for the present month on this blank calendar.

Mon	Tues	Wed	Thurs	Fri	Sat	Sun

➡ *Pupil Book page 121*